T0134919

Studies in Computational Intelligence

Volume 721

About this Series

The series "Studies in Computational Intelligence" (SCI) publishes new developments and advances in the various areas of computational intelligence—quickly and with a high quality. The intent is to cover the theory, applications, and design methods of computational intelligence, as embedded in the fields of engineering, computer science, physics and life sciences, as well as the methodologies behind them. The series contains monographs, lecture notes and edited volumes in computational intelligence spanning the areas of neural networks, connectionist systems, genetic algorithms, evolutionary computation, artificial intelligence, cellular automata, self-organizing systems, soft computing, fuzzy systems, and hybrid intelligent systems. Of particular value to both the contributors and the readership are the short publication timeframe and the worldwide distribution, which enable both wide and rapid dissemination of research output.

More information about this series at http://www.springer.com/series/7092

Roger Lee
Editor

Software Engineering, Artificial Intelligence, Networking and Parallel/Distributed Computing

Editor
Roger Lee
Software Engineering and Information
Technology Institute
Central Michigan University
Mount Pleasant, MI
USA

ISSN 1860-949X ISSN 1860-9503 (electronic)
Studies in Computational Intelligence
ISBN 978-3-319-87221-6 ISBN 978-3-319-62048-0 (eBook)
DOI 10.1007/978-3-319-62048-0

Softcover reprint of the hardcover 1st edition 2017

Printed on acid-free paper

This Springer imprint is published by Springer Nature
The registered company is Springer International Publishing AG
The registered company address is: Gewerbestrasse 11, 6330 Cham, Switzerland

Foreword

The purpose of the 18th IEEE/ACIS International Conference on Software Engineering, Artificial Intelligence, Networking and Parallel/Distributed Computing (SNPD 2017) held during June 26–28, 2017 in Kanazawa, Japan, is aimed at bringing together researchers and scientists, businessmen and entrepreneurs, teachers and students to discuss the numerous fields of computer science, and to share ideas and information in a meaningful way. This publication captures 14 of the conference's most promising papers, and we impatiently await the important contributions that we know these authors will bring to the field.

In Chap. 1, Maoto Inoue, Masato Shirai, and Takao Miura investigate a classification issue of sequence data. They take an approach of probabilistic classification based on Hidden Markov Model (HMM). They build a classifier to each class, apply to sequence data and estimate the class of the maximum likelihood.

In Chap. 2, Marwa Elayni and Farah Jemili present a method to train and combine several datasets from semi-structured sources with the MapReduce programming paradigm under MongoDB. It aims to increase the intrusion detection rates.

In Chap. 3, Yusuke Tanimura, Kazuto Sasai, Gen Kitagata, and Tetsuo Kinoshita propose service-oriented network management with knowledge-based network management support system. The proposed system is modularized and can be applied to fluctuating environment with less burden.

In Chap. 4, Ryotaro Okada, Takafumi Nakanishi, Yuichi Tanaka, Yutaka Ogasawara, and Kazuhiro Ohashi present a dialogue structure analysis method to visualize the transition of topics in a meeting as the one of dialogue process representation. Their method extracts topics in a meeting on time series.

In Chap. 5, Shuangshuang Cai and Mizuho Iwaihara propose a novel embedding method specifically designed for entity disambiguation. Their method jointly maps the information from hierarchical structure of knowledge and context words.

In Chap. 6, Tsukasa Endo, Hasitha Muthumala Waidyasooriya, and Masanori Hariyama propose an automatic optimization method to solve the problem of C-based OpenCL design environment FPGA (field programmable gate array) accelerators.

In Chap. 7, Takeshi Kakimoto, Masateru Tsunoda, and Akito Monden analyze whether team size and duration should be used or not, when they consider the error included in the team size and the duration. As a result, using duration as an independent variable was not very effective in many cases.

In Chap. 8, Prajak Chertchom, Shigeaki Tanimoto, Hayato Ohba, Tsutomu Kohnosu, Toru Kobayashi, Hiroyuki Sato, and Atsushi Kanai present a lifelog attribute data portfolio (LLADP) that will be used for practically modeling life events and for digitizing such information. In this article, they also propose the privacy implications of lifelogging for each attribute.

In Chap. 9, Shota Sakaue, Hiroki Nomiya, and Teruhisa Hochin propose an improved method of obtaining facial expression intensity and estimate emotional scenes based on the basic six facial expressions from lifelog videos.

In Chap. 10, Isao Kikukawa, Chise Aritomi, Shoichi Nakamura, and Youzou Miyadera conduct a study that aims to provide teachers who are contemplating to adopt active learning in existing classes with a framework that facilitates converting existing classes to active learning ones and to realize situation where teachers can instantly convert to active learning.

In Chap. 11, Yumiko Shinohara and Yukiko Nishizaki investigated differences in eye movements, especially fixation duration and location, between novice and expert drivers when driving abroad. The results show the need to develop an automated driving system that considers drivers' driving background.

In Chap. 12, Gongzhu Hu reviews various metrics of password quality, including the one he proposed, and compares their strengths and weaknesses as well as the relationships between these metrics.

In Chap. 13, Kohei Matsumura, Yoshiko Hanada, and Keiko Ono improve the search efficiency of dMSXF by introducing a probabilistic model constructed by the search information to the generation of neighborhood solutions. In the method, the probabilistic model considers nodes individually and a node dependency is ignored.

In Chap. 14, Shingo Takeshita, Takeru Maehara, and Satoshi Ono propose a method for designing a digital watermark that detects replication of 2D code displayed on a smartphone screen. To achieve this, the proposed method designs an effective watermarking scheme for various smartphone models using multi-objective optimization including optical simulation.

It is our sincere hope that this volume provides stimulation and inspiration, and that it will be used as a foundation for works to come.

June 2016

Hiroaki Hirata
Nomiya Hiroki
Kyoto Institute of Technology
Kyoto
Japan

Contents

Contributors

Chise Aritomi Tokoha University, Shizuoka, Japan

Shuangshuang Cai Graduate School of Information, Production and Systems, Waseda University, Kitakyushu, Japan

Prajak Chertchom Thai-Nichi Institute of Technology, Bangkok, Thailand

Marwa Elayni ISITCOM Hammam Sousse, University of Sousse, Sousse, Tunisia

Tsukasa Endo Tohoku University, Sendai, Miyagi, Japan

Yoshiko Hanada Faculty of Engineering Science, Kansai University, Osaka, Japan

Masanori Hariyama Tohoku University, Sendai, Miyagi, Japan

Teruhisa Hochin Information and Human Sciences, Kyoto Institute of Technology, Kyoto, Japan

Gongzhu Hu Department of Computer Science, Central Michigan University, Mount Pleasant, MI, USA

Maoto Inoue Department of Advanced Sciences, HOSEI University, Koganei, Tokyo, Japan

Mizuho Iwaihara Graduate School of Information, Production and Systems, Waseda University, Kitakyushu, Japan

Farah Jemili ISITCOM Hammam Sousse, University of Sousse, Sousse, Tunisia

Takeshi Kakimoto Department of Electrical and Computer Engineering, National Institute of Technology, Kagawa College, Takamatsu, Japan

Atsushi Kanai Hosei University, Tokyo, Japan

Isao Kikukawa Tokoha University, Shizuoka, Japan

Tetsuo Kinoshita Research Institute of Electrical Communication, Tohoku University, Sendai, Japan

Gen Kitagata Research Institute of Electrical Communication, Tohoku University, Sendai, Japan

Toru Kobayashi Nagasaki University, Nagasaki, Japan

Tsutomu Kohnosu Chiba Institute of Technology, Narashino, Japan

Takeru Maehara Department of Information Science and Biomedical Engineering, Graduate School of Science and Engineering, Kagoshima University, Kagoshima, Japan

Kohei Matsumura Graduate School of Science and Engineering, Kansai University, Osaka, Japan

Takao Miura Department of Advanced Sciences, HOSEI University, Koganei, Tokyo, Japan

Youzou Miyadera Tokyo Gakugei University, Tokyo, Japan

Akito Monden Graduate School of Natural Science and Technology, Okayama University, Okayama, Japan

Shoichi Nakamura Fukushima University, Fukushima, Japan

Takafumi Nakanishi Center for Global Communications (GLOCOM) International University of Japan, Tokyo, Japan

Yukiko Nishizaki Information and Human Sciences, Kyoto Institute of Technology, Kyoto, Japan

Hiroki Nomiya Information and Human Sciences, Kyoto Institute of Technology, Kyoto, Japan

Yutaka Ogasawara ITOKI Corporation, Tokyo, Japan

Kazuhiro Ohashi ITOKI Corporation, Tokyo, Japan

Hayato Ohba Chiba Institute of Technology, Narashino, Japan

Ryotaro Okada Center for Global Communications (GLOCOM) International University of Japan, Tokyo, Japan

Keiko Ono Department of Electronics and Informatics, Ryukoku University, Shiga, Japan

Satoshi Ono Department of Information Science and Biomedical Engineering, Graduate School of Science and Engineering, Kagoshima University, Kagoshima, Japan

Shota Sakaue Graduate School of Information Science, Kyoto Institute of Technology, Kyoto, Japan

Kazuto Sasai Research Institute of Electrical Communication, Tohoku University, Sendai, Japan

Hiroyuki Sato The University of Tokyo, Tokyo, Japan

Yumiko Shinohara Graduate School of Information Science, Kyoto Institute of Technology, Kyoto, Japan

Masato Shirai Department of Mathematics and Computer Science, Interdisciplinary Faculty of Science and Engineering, Shimane University, Matsue, Shimane, Japan

Shingo Takeshita Department of Information Science and Biomedical Engineering, Graduate School of Science and Engineering, Kagoshima University, Kagoshima, Japan

Yuichi Tanaka ITOKI Corporation, Tokyo, Japan

Shigeaki Tanimoto Chiba Institute of Technology, Narashino, Japan

Yusuke Tanimura Graduate School of Information Sciences, Tohoku University, Sendai, Japan

Masateru Tsunoda Department of Informatics, Kindai University, Higashiosaka, Japan

Hasitha Muthumala Waidyasooriya Tohoku University, Sendai, Miyagi, Japan

Sequence Classification Based on Active Learning

Maoto Inoue, Masato Shirai and Takao Miura

Abstract In this investigation, we discuss a classification issue of sequence data. Generally, we assume a set of training data to construct classifiers, but the construction is not easy to obtain such data set. We take an approach of probabilistic classification based on Hidden Markov Model (HMM). We build a classifier to each class, apply to sequence data and estimate the class of the maximum likelihood. HMM requires less amount of training data but these data help HMM to work better. We propose an *active learning* approach to construct classifiers. The basic idea is that HMM takes a new training data autonomously to polish up the classifiers whenever HMM expects the more likelihood.

Keywords Sequence classification · Hidden Markov Model · Active learning

1 Introduction

Nowadays there generate a huge amount of information and we can get to them easily and quickly through internet. A *data mining* helps us to extract useful knowledge from these information; this is one of the current topics of research interests. However, it is not easy to do the work because of the huge amount of various data, and very often they disappear immediately. Certainly we need efficient, accurate and useful techniques to extract knowledge.

M. Inoue (✉) · T. Miura
Department of Advanced Sciences, HOSEI University, Kajinocho 3-7-2, Koganei, Tokyo 186–8584, Japan
e-mail: maoto.inoue.7h@stu.hosei.ac.jp

T. Miura
e-mail: miurat@k.hosei.ac.jp

M. Shirai
Department of Mathematics and Computer Science, Interdisciplinary Faculty of Science and Engineering, Shimane University, 1060 Nishi-Kawazu, Matsue, Shimane 690–8504, Japan
e-mail: shirai@cis.shimane-u.ac.jp

R. Lee (ed.), *Software Engineering, Artificial Intelligence, Networking and Parallel/Distributed Computing*, Studies in Computational Intelligence 721,
DOI 10.1007/978-3-319-62048-0_1

In this work, we discuss classification of sequence data. Unlike simple structured data or formatted data, there may arise sequence data very often, such as sentences, melodies and program codes in internet. We assume a sequence consists of elements (possibly with labels) and these labels play their own roles depending upon the class.

To tackle with the issue of sequence data, there have been some sophisticated tools based on probabilistic and stochastic theory proposed so far such as *Hidden Markov Model* (HMM), Maximum Entropy Markov Model (MEMM) and Conditional Random Fields (CRF). HMM requires less amount of data to estimate models but more data help HMM to work better. Although MEMM and CRF are powerful to model sequences, we should have some amount of training data in advance.

However, what we mean by training data? How we can put labels on (parts/whole of) sequences? To what extent we should assume domain-specific knowledge? We can avoid subjectivity for training? More seriously, it is not easy to obtain training data, it takes much time/cost to complete the work, and we hardly get *many* training data.

In this investigation, we propose a new approach of sequence classification based on HMM using active learning (called *Active Learn HMM*, ALHMM).

This work contributes to the following 3 points:

(1) We can construct necessary data for better construction.
(2) We choose data small enough to construct reliable classifiers.
(3) We implement efficient algorithm of ALHMM for model construction.

The rest of the paper is organized as follows. In Sect. 2, we define the classification and training data. In Sect. 3, discusses a background of HMM and Active Learning. In Sect. 4, we propose our approach of ALHMM and in Sect. 5 we show experimental results to see how effective our approach works. In Sect. 6, we conclude this investigation.

2 Classification and Training Data

Classification allows us to put information into one of the distinct groups given in advance. We call such group a *class* (with a *label*) and the rules to classify are called *classifier*. However, there exist a vast amount of information and it takes much time/cost to classify them. More seriously, the classification results may depend on subjectivity.

To overcome these problems, we introduce a notion of *classifier* which is a system to classify information automatically. Such a system provides us with quick, cheap and efficient processing.

By examining information content, we see several *features* to characterize, identify and classify information. A class *baseball* could be identified by feature words such as `"pitcher"`, `"Ichiro"`, `"home-run"`. Once we find these features, we expect to classify *baseball* correctly by examining whether the features appear in the information or not.

How can we extract features to classify data? Generally we assume a set training data to construct classifiers [8], such an approach is called *supervised learning*. The training data provides us with some knowledge how to examine data and to decide classes. Decision tree, k-NN and several Bayesian approach have been proposed so far [8]. But clearly we face to two problems. Classifiers require many training data and also they assume *simple* and *formatted* data.

Basically the more training data we have, the better classifier we find. Thus, we need much amount of training data and much more time to do that. This causes some problems; it's hard to follow heavy changes (called *burst* situation) and hard to mind structured information such as sequence.

As for sequence classification, there have been proposed some approaches such as document classification based on N-gram. By putting consecutive N words into one (called N-gram), we could examine frequency, co-occurrence and relationship among them. However we may have large but sparse distributions of N-grams which cause heavy complexity of space and computation [8].

Here we take a probabilistic approach based on HMM for classification of sequence data. For this purpose, we construct a classifier to each class and apply sequence data. It is not straightforward to apply well-known classification techniques to sequences because they are lack of *sequence* concept, and we estimate the class by means of probabilistic tools and maximum likelihood.

3 HMM and Active Learning

In this section, we review some background knowledge and our motivation of the approach.

3.1 Markov Model

In probability theory, a *stochastic process* means a mathematical tool where random variables are associated with a set of states to model randomly changing over time, considered as a probabilistic finite state automaton (with observation output). The approach is widely used as mathematical models of systems and phenomena that appear to vary in a random manner. Common example is "speech recognition".

A *Markov Model* (MM) is one of the stochastic framework with an assumption that future states depend only on the current state but not on the events/states that occurred before it, called *Markov property*.

Generally, this assumption enables reasoning and computation to intractable systems, and facilitates predictive modelling and probabilistic forecasting in a simple and efficient way [9].

However, MM has 2 serious deficiencies: how to estimate *which* state we stand on and how to build training data. Usually we could construct the model by examining training data, say counting state transitions and observation-output. It can be noted that there exist two kinds of *labels*, one for a state and another for class. Clearly the true problem is how we obtain many training data correctly and efficiently.

3.2 HMM and Classification

A *Hidden Markov Model* is a kind of Markov Model for which the state is only partially observable [3]. Note observation-outputs are related to the states of the system, but they are typically insufficient to precisely determine the state. What *hidden* does mean is a *state* associated with observation for *best* suitable interpretation.

HMM provides us with solutions to 3 issues. Assume we have observation sequences $O = o_1, \ldots, o_N$. The first is that we can obtain the probability of O. In fact, once we assume a Markov Model of interests, we explore all the paths to generate O, we summarize all the probabilities of the paths. The second is that we can obtain the *most likely* path to generate O. *Viterbi* algorithm allows us to estimate the path of the maximum likelihood efficiently using dynamic programming technique: The forward algorithm computes the probability of the sequence of observations.

One important application is that we can consider HMM as a classifier of sequence data using *Maximum Likelihood Principle* (MLP). That means, when we construct MM to each class and we obtain the likelihood of the sequence (by Viterbi), we decide a class of the sequence of the maximum.

For example, a *Sonata* class carries a musical structure consisting of 4 main sections: an *exposition* of theme, a *development*, a *recapitulation* and *coda*. The form appears widely after many music of 18th century (the early Classical period) [*WIKI*]. Once every part of music may be classified into one of the sections, we could say the music has sonata form, thus part of music as well as their labels constitute a class "Sonata".

Given *Sonata* Markov Model and a music score, we estimate the likelihood to the state transition of an *exposition*, a *development*, a *recapitulation* and *coda* segmenting the music con considering them as observation-output.

Similarly we obtain the values for *Rondo*, *Ballade* and so on. Then we classify the music form as the one of the maximum.

True contribution of HMM is that we can estimate Markov Model, i.e., state transition probabilities $P(s_j|s_i)$, probabilities $P(o_k|s_i)$ of observation o_k at state s_i and initial probability $P(s_i)$ at start. In other words, HMM generates MM according to sequence data.

Basically we apply EM algorithm to estimate MM. We maximize the likelihood of the data by iterating adjustment process through EM algorithm from scratch. HMM requires *no training data* in advance. Several well-known algorithms for Hidden Markov Models exist. *Baum Welch* (BW) algorithm estimates them efficiently from a set of (untrained) data.

However, a sufficient size of "training data" helps us to start with better initialization suitable for quick convergence (of the iteration) and for avoidance of local-minimum situation. Without any training data or appropriate initialization, it is hard to avoid these deficiencies.

Here we introduce a notion of *development data* instead of *training* data. Each development data contains label for class but not labels for states. Although it takes time and cost to construct training data to HMM, we could obtain development data much easier. For example, we know several articles of sport-news have a label *sport* and handful sonata music. Important is MLP works well with development data. In this investigation, we stick to this aspect and improve the relevant parameters.

3.3 Active Learning

Active Learning is a framework of automatic and autonomous selection of data *useful* for training to achieve high accuracy of classification. Here by a word "useful", we mean the data to improve precision of classifiers. Beginning with a few data, we could expect more training data whenever we need to learn more. Here what we have to pay attention is under what conditions we can decide to select these data.

There exist at least 2 advantages in Active Learning [6]. The first one is that it is possible to achieve high accuracy of classification without large training data in advance. The second is that the results don't depend on annotators subjectivity, i.e., we can put common, reasonable and integrated conditions to construct training data.

There have been several investigation of active learning so far where 3 *scenarios* have been proposed in which we could ask unlabeled data of being labeled. The first scenario is *pool-based* [10]. We select some data from large unlabeled *data pool* where data are asked in a greedy fashion according to an *informativeness* measure to evaluate all instances in the pool.

The second is called *stream-based* [4, 5] where, whenever a new data comes in, we are asked to be included in the training data or not. Here we assume it is inexpensive to obtain an unlabeled instance and we are asked to decide whether or not to request its label. The decision can be framed several ways such as using some informativeness measure or any other machine learning techniques like conceptual clustering.

The last one is *membership query syntheses* [2] which means we can request labels for unlabeled data in input stream. There could exist human annotators to label such arbitrary data, and can be awkward. Here we will take this approach to improve HMM.

Active learning scenarios involve evaluating the informativeness of unlabeled data. There have been many proposal of formulating strategies, such as *Uncertainty Sampling, Query By Committee, Expected Model Change* and so on [11]. Among others, we put our attention on 3 strategies based on "uncertainty sampling" scenario, *Margin Sampling* [12], *Entropy-based Approach* and *Least/Largest Confidence* [7].

Margin Sampling means we select a data of the largest data probability which has the minimum difference between the largest and the second largest probabilities. By *entropy-based approach*, we select a new data by which we may keep the least entropy with in the class. Finally *Least/Largest Confidence* provides us with a new data of the maximum which has the minimum probability of class membership (or the minimum of the maximum membership).

On the other hand, we could have at least two disadvantages of the construction of training data. The first one is so-called *sampling bias* issue, i.e., when we take samples according to some conditions, they might not always play best. Possible solution is, for example, *importance sampling* where we specialize topics of interests for sampling. The second disadvantage is that we utilize identical data many times for training because we might change parameters in conditions.

4 Active Learning for HMM

In this section we propose a new approach to combine HMM and Active Learning, or *Active Learning for HMM*. The approach is not so clear because we don't know yet class-membership probability during the iteration of HMM estimation enough to decide selection of data.

Anderson et al. have discussed the estimation of HMM parameters, state transition and the best states in terms of Active Learning [1]. They involved *loss-functions* which tell us to what parts we should improve the model using training data. They have shown the effective of less-training and more-efficient situation compared to uncertainty sampling. But they don't discuss model likelihood (and the classification).

Here we propose model calculation, called *ALHMM* of HMM using Active Learning. In a naive HMM approach, we don't assume any training data (data with state labels) but examine just the likelihood of sequences. We maximize the value until the saturation based on EM algorithm. Here it is enough to discriminate class membership no matter how big specific likelihood is.

Here we assume there exist a collection of test sequences to be classified and another collection of model development sequences to each class. According to Membership-Query-Synthesis scenario, we show out ALHMM algorithm for HMM estimation:

[0]: Let $\rho = 0.5$. For each new sequence do [1] to [4].
[1]: We estimate new HMM parameters to each class from the current parameters.
[2]: We initiate the decision whether MLP can be applied or not to see the best likelihood goes over a threshold ρ compared to the second likelihood.
[3]: If not, we select a new sequence and its class for better model construction from development data, and estimate yet new HMM parameters by the data to the class. Note the new sequence (development data) has a class label without any state labels and may help us to improve the class-likelihood.
[4]: We apply MPL and decide the class for the new sequence. Also we keep improving ρ dynamically. That is, if the (new) best likelihood is below a threshold ρ compared to the (new) second likelihood, we decrease ρ by multiplying 0.1. If the (new) best likelihood is

above a threshold ρ compared to the (new) second likelihood, we decrease ρ by multiplying 10.0.

In ALHMM (3), we select one development sequence if and only if (a) the sequence has the length less than σ, and (b) the sequence satisfies Least Confidence or Largest Confidence.

We give σ as an average size of all the development data. Note generally the shorter sequence may have larger likelihood, this means a new development data may allow us to improve average likelihood of the class.

When we adopt Least Confidence principle, we can polish up membership probability of cloudy classes, while Largest Confidence principle helps us to avoid small classes and to polish up class sizes.

5 Experiments

5.1 Preliminaries

In this section we show how well ALHMM does work for sequence classification. We examine the proposed approach an ALHMM with Largest Confidence (called *ALHMM)*) and a (naive) HMM as a baseline.

We examine 465 news articles of 3 classes in `Daily Yomiuri` 2007 January in English (from the beginning) for our experiments and examine precision/recall factors as well as harmonic averages (f-values). Assuming *sport, economics* and *science* classes, we start with identical 15 articles (5 articles $\times$ 3 classes = 15) as initial data to ALHMM and HMM. To compare ALHMM with HMM, we also examine HMM with development data sets of $15(5 \times 3), 21(7 \times 3), 27(9 \times 3), 30(10 \times 3), 36(12 \times 3), 42(14 \times 3)$ and $45(15 \times 3)$ articles. Note we examine the same sets of development data and test data.

We apply GoTagger as morphological processing to the corpus and extract only nouns and verbs in advance. As initial data, development data for ALHMM and test data, we examine 45 articles as above, $90 \times 3 = 270$ articles with an average size $\sigma = 15.93$ words and $50 \times 3 = 150$ articles respectively.

Table 1 contains some statistics of first 8 articles in each class.

Let us illustrate an article with tags in a Fig. 1.

Every article in the corpus carries a class label by which we may select appropriately.

As we said, we give $\sigma = 15.93$ words that is an average size of all the development data. Among 270 development data, there exist 130 articles in total (59, 41 and 30 articles of class *sport, science* and *economics* respectively) whose size are above σ.

As an evaluation measure, we examine f-values to all the 150 test articles as well as micro precision and recall.

Table 1 Development articles

sport		science		economics	
ID	Size	ID	Size	ID	Size
1	8	101	19	201	23
2	16	102	15	202	9
3	21	103	16	203	22
4	15	104	6	204	18
5	13	105	8	205	19
6	7	106	26	206	23
7	12	107	16	207	21
8	12	108	21	208	24

```
1 winter_NN␣Japan_NNP␣teams_NNS␣begin_VBP␣quest_NN␣league_NN␣titles_NNS␣Japan_NNP␣Series_NNP␣rings_NNS␣today_NN␣spring_NN␣training_NN␣opens_VBZ␣Kyushu_NNP␣Okinawa_NNP␣Australia_NNP↓
2 chance_NN␣address_VB␣needs_NNS␣BayStars_NNP␣loaded_VBD␣arms_NNS␣ignored_VBN␣bats_NNS␣missed_VBD␣boat_NN↓
3 Hingis_NNP␣showed_VBD␣developed_VBN␣game_NN␣thumping_VBG␣Australia_NNP␣Nicole_NNP␣Pratt_NNP␣advance_VB␣quarterfinals_NNS␣dollars_NNS␣Toray_NNP␣Pan_NNP␣Pacific_NNP␣Open_NNP↓
4 coach_NN␣Top_NNP␣League_NNP␣team_NN␣Mitsubishi_NNP␣Juko_NNP␣Scott_NNP␣Pierce_NNP␣is_VBZ␣going_VBG␣get_VB␣opportunities_NNS␣play_VB↓
5 Ai_VBP␣Sugiyama_NNP␣confirmed_VBD␣Thursday_NNP␣was_VBD␣looking_VBG␣doubles_NNS␣partner_NN␣replace_VB␣Daniela_NNP␣Hantuchova_NNP␣has_VBZ␣played_VBN␣May_NNP↓
6 New_NNP␣York_NNP␣Yankees_NNP␣have_VBP␣come_VBN␣Asia_NNP␣build_VB␣game_NN␣end_NN␣turn_VB␣profit_NN↓
7 Sugiyama_NNP␣made_VBD␣adjustment_NN␣needed_VBD␣rallied_VBD␣defeat_VB␣Russia_NNP␣Maria_NNP␣Kirilenko_NNP␣advance_VB␣Thursday_NNP␣quarterfinals_NNS␣dollars_NNS␣Toray_NNP␣Pan_NNP␣Pacific_NNP␣Open_NNP↓
```

Fig. 1 A sample article

5.2 Results

Here are our summary of the classification in a Table 2.

As the Table 2 says, the more development data we have the better f-values we find in HMM. In fact, starting with 0.391 of HMM with 15 articles (called HMM15), we get to 0.672 in HMM45.

ALHMM, starting with 15 articles, requires 12 articles additionally during the process, gets to 71.3% precision and 0.619 f-value. The result is comparable to HMM42, i.e., we obtain similar result with 27/42 = 60% articles by ALHMM. Because HMM15 shows 47.3% precision and the f-value 0.391, ALHMM gives us the excellent improvement of 0.619/0.391 = +58.3% in f-value (+51.3% precision).

Let us show how ALHMM works in a Table 3. Here we see additional 12 articles which contain 4 science articles, 6 economics and 2 sport.

To compare ALHMM with others, we show the detail of HMM15, HMM45 and ALHMM in a Table 4. As expected, HMM45 (75.3%) works much better than HMM15 (47.3%) while ALHMM (15 + 12) shows the comparable one (71.3%).

Table 2 Classification results

Articles	Precision	Recall	f-value
(HMM)			
15	0.473	0.333	0.391
21	0.593	0.473	0.527
27	0.62	0.42	0.501
30	0.647	0.413	0.504
36	0.7	0.487	0.574
42	0.707	0.46	0.557
45	0.753	0.607	0.672
(ALHMM)			
15 + 12	0.713	0.547	0.619

Table 3 Active learning in ALHMM

Iteration	ρ	Failed TestArticle	Selected Class	Development Article
1	0.5	2	science	102
2	$0.5 \rightarrow 0.1$	2, 3	science	116
3	$0.1 \rightarrow 0.01$	3, 7	economics	236
4	$0.01 \rightarrow 0.0001, 0.0001 \rightarrow 0.01$	7, 20	economics	251
5	$0.01 \rightarrow 0.0001, 0.0001 \rightarrow 0.01$	20, 32	sport	54
6	$0.01 \rightarrow 0.0001, 0.0001 \rightarrow 0.01$	32, 46	economics	230
7	$0.01 \rightarrow 0.0001, 0.0001 \rightarrow 0.01$	46, 62	economics	244
8	$0.01 \rightarrow 0.0001, 0.0001 \rightarrow 0.01$	62, 80	sport	14
9	$0.01 \rightarrow 0.0001, 0.0001 \rightarrow 0.01$	80, 91	science	109
10	$0.01 \rightarrow 0.0001, 0.0001 \rightarrow 0.01$	91, 102	economics	241
11	$0.01 \rightarrow 0.0001$, $0.0001 \rightarrow 0.01 \rightarrow 0.1 \rightarrow 0.5$	102, 133	science	147
12	$0.5 \rightarrow 0.1$	133, 134	economics	261
13	$0.1 \rightarrow 0.01, 0.01 \rightarrow 0.1$	134, –	(None)	

We see that ALHMM outperforms HMM15 in all of sport (+4.0%), science (+10.0%) and economics (+50%) classes, and is comparable with HMM45.

Figure 2, Fig. 3, Fig. 4 for ALHMM and Fig. 5, Fig. 6, Fig. 7 for HMM45 contain the Markov Models of ALHMM and HMM45 respectively where circles show states, arrows describe transition between them, squares contain frequent observation words depending on their states. Names within the circles mean state-labels where states of the same names share the several observation words.

We obtain exactly same initial states of the biggest probabilities in 6 models as shown in a Table 5: Player in "sport" class, Organ in "economics" and Research in "science" for ALHMM, HMM45 and HMM15.

Table 4 Micro precision (HMM)

Corrects/Answer	sport	science	economics	Unknown	Total
(HMM15)					
sport	31	2	2	15	50
science	21	6	10	13	50
economics	18	11	13	8	50
(HMM45)					
sport	30	9	11	0	50
science	10	26	14	0	50
economics	7	8	35	0	50
(ALHMM)					
sport	33	3	14	0	50
science	19	11	20	0	50
economics	7	5	38	0	50

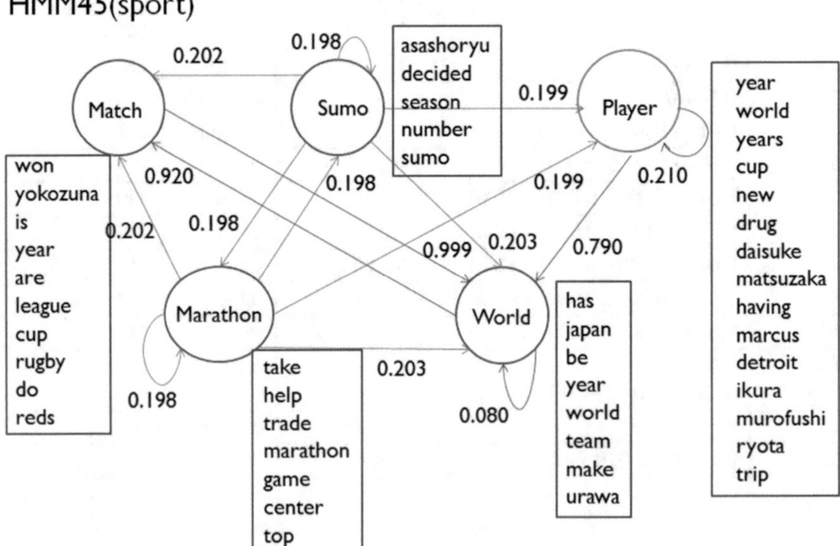

Fig. 2 "sport" Markov Models: HMM45

It is worth noting that the most likely paths in "economics" and "science" classes of ALHMM and HMM45 become identical as shown in a Table 6, while "sport" has slightly different one.

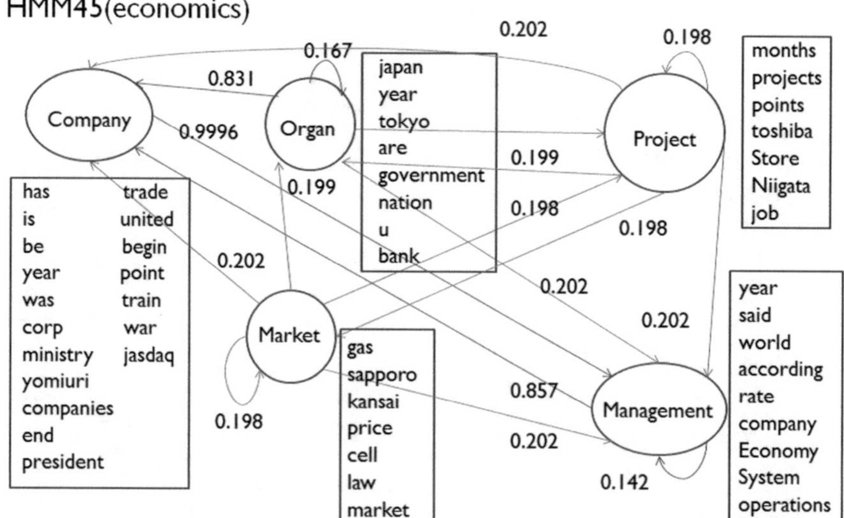

Fig. 3 "economics" Markov Models: HMM45

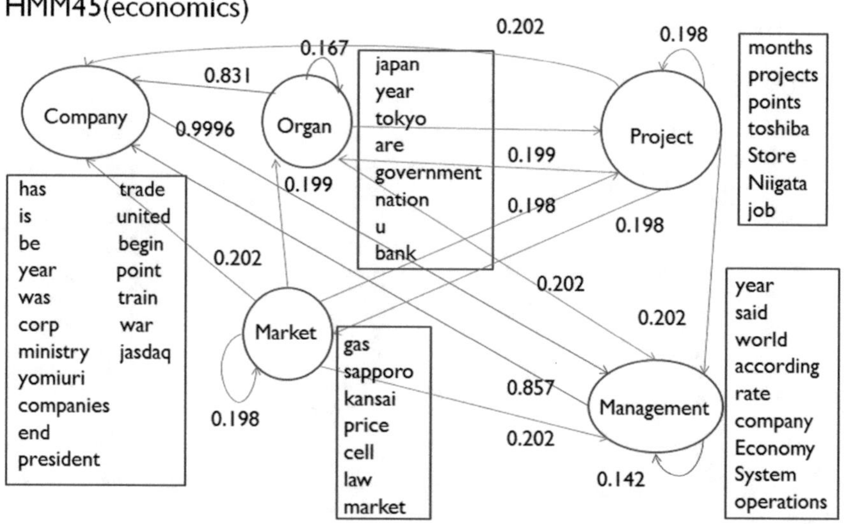

Fig. 4 "science" Markov Models: HMM45

5.3 Discussion

Clearly we can say ALHMM outperforms HMM by looking at precision, recall and f-values with smaller amount of data. In fact, ALHMM shows the results similar to

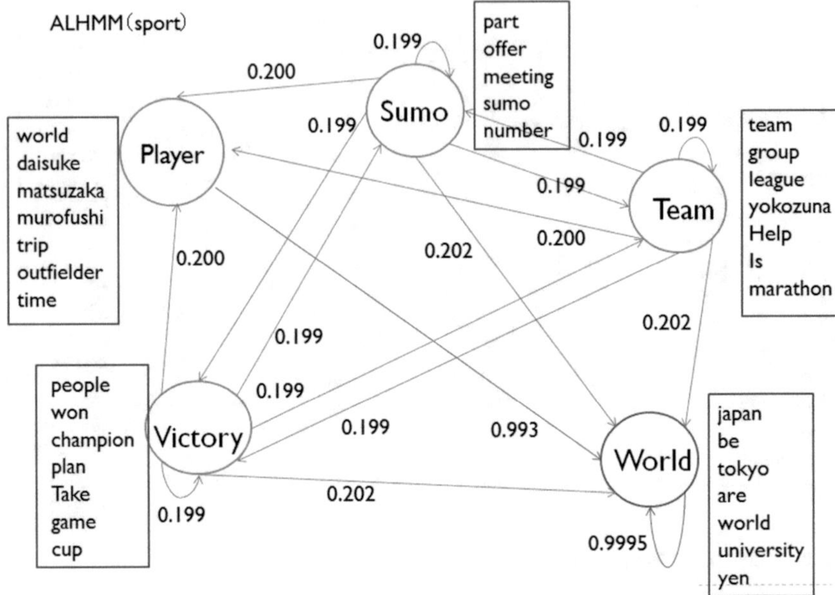

Fig. 5 "sport" Markov Models: ALHMM

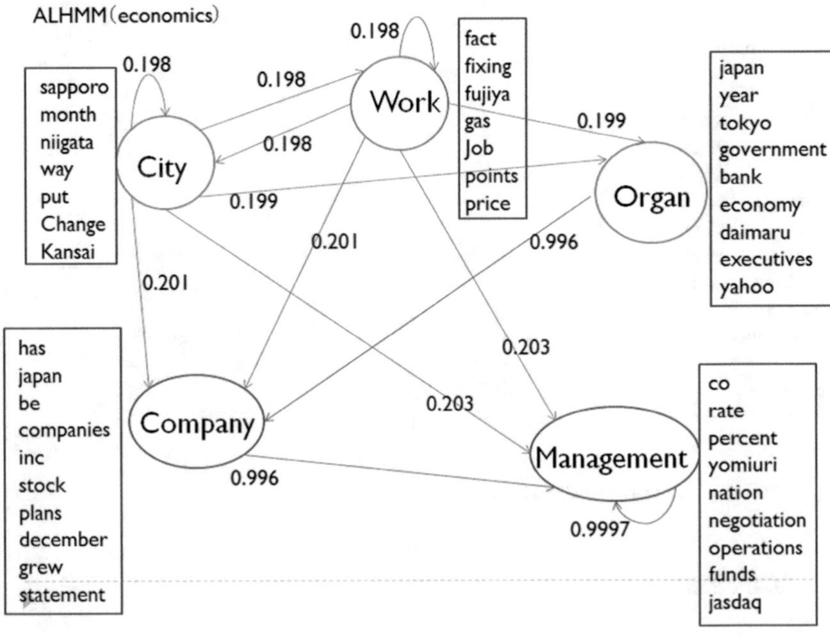

Fig. 6 "economics" Markov Models: ALHMM

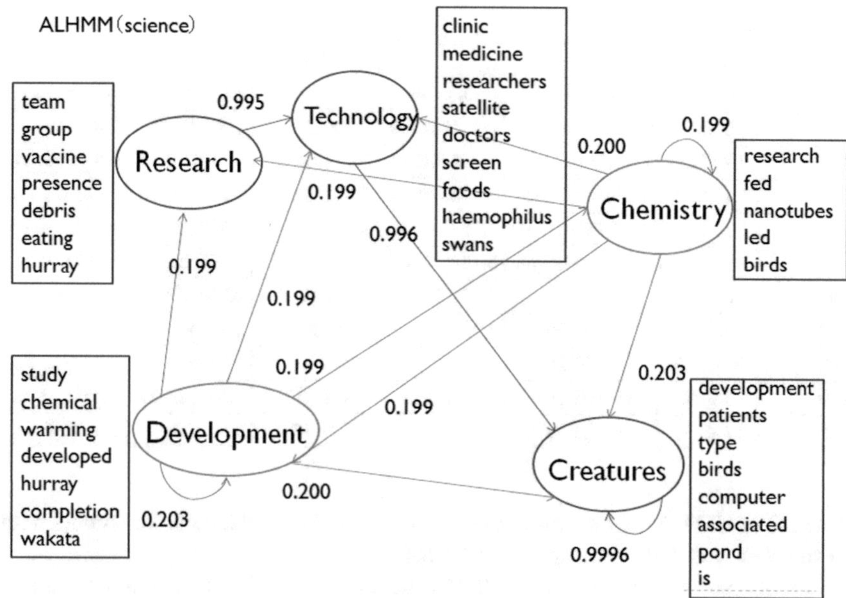

Fig. 7 "science" Markov Models: ALHMM

Table 5 Initial states and probabilities

	sport	economics	science
HMM15	Player	Organ	Research
	0.991	0.992	0.992
HMM45	Player	Organ	Research
	0.997	0.9996	0.997
ALHMM	Player	Organ	Research
	0.994	0.996	0.996

Table 6 Most likely paths

class	ALHMM/HMM45
sport	Player → World → World → …
	Player → World → Match → World → Match → …
economics	Organ → Company → Management → Management → …
	Organ → Company → Management → Management → …
science	Research → Technology → Creatures → Creatures → …
	Research → Technology → Creatures → Creatures → …

Table 7 Likelihood of articles

ID	HMM15			ALHMM		
	sport	science	economics	sport	science	economics
101	4.00817E-47	9.69406E-52	4.71131E-49	1.75021E-50	1.3714E-50	2.4086E-48
102	9.72295E-34	9.75243E-34	9.75243E-34	9.78194E-34	9.81927E-34	<u>9.85679E-34</u>
103	1.16909E-26	3.81107E-26	2.83387E-26	9.80153E-28	9.82713E-28	2.46568E-26
104	1.16675E-32	9.75243E-34	9.75243E-34	9.78194E-34	9.81927E-34	2.4642E-32
105	2.39051E-64	3.75806E-68	3.12208E-67	9.6652E-70	3.96753E-67	1.69713E-66
106	1.66E-75	3.75E-77	1.73E-77	9.63623E-79	2.83055E-77	1.413E-76
107	9.59E-76	9.62E-76	1.73E-74	9.64588E-76	9.76443E-76	1.17786E-74
108	9.74E-28	9.77E-28	9.77E-28	9.80153E-28	9.84689E-28	<u>9.86271E-28</u>
109	9.76E-22	9.79E-22	9.79E-22	9.82117E-22	9.85478E-22	<u>9.86863E-22</u>
110	1.16E-56	9.67E-58	3.13E-55	1.74671E-56	9.80758E-58	2.45829E-56

HMM42 or HMM45, which mean the fact that we get equivalent markov model with 40 and 36% less data by means of ALHMM.

In fact, starting with the identical 15 articles, ALHMM always explore markov models in a sense of better likelihood. We see that unknown test articles in HMM15 can be estimated correctly in ALHMM. A Table 7 contains the likelihood values in these cases:

To see why ALHMM and HMM45 play similar performance with each other let us compare ALHMM (Figs. 2, 3 and 4) with HMM45 (Figs. 5, 6 and 7). As in a Table 6, we see exactly same paths that are the most likely in "economics" and "science" classes. This means we have similar interpretation in both HMM45 and ALHMM. There happens some difference in "sport" class (Figs. 2 and 5).

Two states Player and World in "sport" class of HMM45 and ALHMM share several observation words so that they look like similar. A state Match in "sport" of HMM45 also shares some words ("year") in a state World. Then we apply mathematical test of goodness of fit to word distribution of these states.

In World/Match of HMM45 there happen 14 observation words {`is, year, are, league, cup, do, reds, has, japan, be, world, team, make, urawa`}. Also in World of ALHMM/ALHMM we have 12 observation words {`japan, be, tokyo, are, world, university, yen, has, year, team, make, urawa`}. In the first case, X^2 value is 3.522982 and χ^2-value is 22.36 with the significance 5%. In the second case, X^2 value is 2.520914 and χ^2 value is 19.68. Both cases say World and Match play same role in our estimation.

6 Conclusion

In this work, we have proposed sequence classification based on Hidden Markov Model (HMM) approach. Since it is hard to prepare training data appropriately for the task, we have proposed ALHMM which takes Active Learning approach to improve likelihood of models. In this approach we have constructed a Markov Model to each class and have selected candidate date to make discrimination more sharp.

In our experimental results, we have shown that we get the models similar to naive models with 40% less amount of information while keeping comparable precision and recall.

References

1. Anderson, B., Moore, A.: Active learning for Hidden Markov Models: objective functions and algorithms. In: Proceedings of ICML (2005)
2. Angluin, D.: Queries and concept learning. Mach. Learn. **2**, 319–342 (1988)
3. Bilmes, J.A.: A gentle tutorial of the EM algorithm and its application to parameter estimation for Gaussian Mixture and Hidden Markov Models. In: Proceedings of ICSI (1998)
4. Bouguelia, M.R., Belad, Y., Belad, A.: A stream-based semi-supervised active learning approach for document classification. In: Proceedings of ICDAR (2013)
5. Cohn, D., Atlas, L., Ladnar, R.: Improving generalization with active learning. Mach. Learn. **15**(2), 201–221 (1994)
6. Dasgupta, S., Langford, J.: A tutorial on active learning. In: Proceedings of ICML (2009)
7. Dredze, M., Crammer, K.: Active learning with confidence. In: ACL08, pp. 233–236 (2008)
8. Han, J., Kamber, M., Pei, J.: Data Mining: Concepts and Techniques, 3rd edn. Morgan Kaufmann (2011)
9. Jurafsky, D., Martin, J.H.: Hidden Markov Model. Speech and Language Processing (2016)
10. Lewis, D., Gale, W.: A sequential algorithm for training text classifiers. In: Proceedings of ACM SIGIR Conference on Research and Development in Information Retrieval, pp. 3–12 (1994)
11. Settles, B.: Active learning literature survey. In: Proceedings of Computer Sciences, Technical Report 1648, University of Wisconsin Madison (2010)
12. Zhou, J., Sun, S.: Improved margin sampling for active learning. Proc. Commun. Comput. Inf. Sci. **483**, 120–129 (2014)

Using MongoDB Databases for Training and Combining Intrusion Detection Datasets

Marwa Elayni and Farah Jemili

Abstract A single source of intrusion detection dataset involves the analyze of Big Data, recent attempts focus on Big Data techniques in order to combine heterogeneous data sets and solve the problems of analyzing the huge amounts of data. The main objective of this paper is to present a method to train and combine several datasets from semi-structured sources with the MapReduce programming paradigm under MongoDB. It aims to increase the intrusion detection rates. In our work, we will focus on KDD99, DARPA 1998 and DARPA 1999 dataset and with the big data technique MapReduce in MongoDB: First, we will select the most pertinent attributes and eliminate the redundancies from the previous datasets. Then, we will merge them vertically into the same collection. Finally, to analyze the dataset we will use a Bayesian network as K2 algorithm implemented in WEKA.

Keywords Intrusion detection · Big Heterogeneous Data · NOSQL system · MapReduce · MongoDB · KDD99 · DARPA · K2

1 Introduction

All intrusion detection systems involve an intrusion detection dataset, a learning system and an inference system. Most existing researches deploy a single intrusion detection dataset for system learning and inference for which these works use several classification methods such as Bayes networks, neural networks, fuzzy neural network and genetic algorithms [1–4].

The results of the intrusion detection of these works do not show high performance at intrusion detection rate, so the idea is to invest within the intrusion detection datasets to increase the intrusion detection rates.

M. Elayni (✉) · F. Jemili (✉)
ISITCOM Hammam Sousse, University of Sousse, Sousse, Tunisia
e-mail: aynimarwa@gmail.com

F. Jemili
e-mail: Jmili_farah@yahoo.fr

R. Lee (ed.), *Software Engineering, Artificial Intelligence, Networking and Parallel/Distributed Computing*, Studies in Computational Intelligence 721,
DOI 10.1007/978-3-319-62048-0_2

In literature we find many of intrusion detection datasets such as KDD 99, NSL KDD, DARPA, CAIDA, ADFA,… these datasets are characterized by a huge data size, unstructured format. Then a use of Big Data techniques is essential in the intrusion detection. And now due to Big Data, we are able to manage and treat a huge size of instances stored in different intrusion detection datasets.

In this context, we cite the example of the commercial SIEM (Security Information and Event Management) [5, 6] using relational database technologies for storage repositories, found that databases has become bottlenecks in deployments at larger enterprises: storage and retrieval of data start to taken more time which is unacceptable.

It is a clear that SIEM data, presenting the intrusion detection traffic, are facing Big Data challenges where as the relational databases are becoming bottlenecks.

So, next generation Big Data storage technologies like NoSQL databases can help in addressing these problems.

The purpose of this paper is to describe a training model for three datasets and to fuse those datasets in a single dataset with NoSQL database MongoDB to achieve the goal which is obtaining higher intrusion detection rates and lower false alarms.

We started by introducing the work. The second section describes the intrusion detection datasets used in our model. In the third section, we justify our choice of MongoDB as NoSQL system. In the fourth section we explain the use of MongoDB. In the fifth section, we provide experimental results.

2 Intrusion Detection Datasets

In our work we use three datasets KDD99, DARPA98 and DARPA99, our choice is based on a study by Azad and Jha [7] published in a journal "the intrusion detection and Big Heterogeneous Data" [6]: 46 out of 75 studies used either DARPA or KDD Cup while only 29 chose a different one.

Although these two dataset have been there for over a decade, they are still considered as the two most popular datasets used for intrusion detection researches. Even in our approach, we used the following datasets:

KDD99

The KDD Cup 1999 dataset which is used in our experiment [8], used for benchmarking intrusion detection problem. The dataset is a collection of simulated raw TCP dump data during a period of nine weeks on a local network area: seven weeks of network traffic that gives about five million connections records of the training data and two weeks of testing data is giving around two million connections records.

DARPA

DARPA99 and DARPA 98 traces are generated by the MIT Lincoln Labs for intrusion detection evaluation [9, 10]. The DARPA98 traces are consisted of: training data seven weeks and testing data two weeks.

Table 1 The Name of some attacks of each category _KDD, DARPA

Categories	Name of attack
Probing	Ipsweep, nmap, portsweep, satan, IPsweep, saintmscan, nmap, …
DOS	Neptune, pod, land, back, smurf, teardrop, …
U2R	Loadmodule, buffer_overflow, rootkit, perl, format, PS, …
R2L	Imap, ftp_write, Warezclient, multihop, phf, spy, guess_passwd, warezmaster, imap, worm, …

The DARPA 99 is consisted of five weeks: the first three weeks are dedicated for training data, while the last two weeks are for testing and that gives about six million connections records.

For each connection present in KDD and DARPA, there are attributes which are dedicated for each dataset, these attributes describe different features of the connection and contain a specific label assigned to indicate the type of label and whether it's an attack or normal one. There is a large variety of attacks; most of them are grouped into one of the following categories [11]:

- A Probe which is an attempt to learn information that could facilitate an attack.
- A Denial of Service (DoS) which is an attack that overloads the resources of a system and aims to make the services or the resources of an organization unavailable during an indefinite time.
- A user to root (U2R) which is an attack where a user with limited right access attempts to gain root permissions.
- A remote to user (R2L) it's the case when a user, that is unknown to the system, attempt to own the user permissions in order to expose a machine vulnerabilities via internet.

The Table 1 shows some attacks present in KDD and DARPA and for each attack we indicate its category.

In order to present our model for training and fusion those three datasets, we used NoSQL technique, in the next section; we define the NoSQL systems, ending in approving our choice of the best efficient NoSQL systems.

3 NoSQL Databases

Since 2012 the data volume has evolved from a few dozen terabytes to many petabytes [12], so necessary techniques are mandatory to capture, manage and process these huge amounts of datasets "Big Data" within a tolerable elapsed time.

Currently, there are several solutions for the analysis of big data that can be mentioned: Search-based systems, New SQL and NoSQL databases [12]. We are interested in NoSQL databases for resolving the problem of analytic data for intrusion detection system.

Table 2 Types and examples of NOSQL databases

Types	Examples
Key/value databases	Couchbase, Dynamo, redis, Riak, orientDB…
Document store	MongoDB, Apache CouchDB…
Column-oriented	Cassandra, HBase, vertica…
Graph	Neo4J, Allegro Graph, stardog…

There are many types of data storage models in NoSQL databases that are classified into four categories. In Table 2 we present examples of NoSQL databases for each category.

In this context, authors in [12] have compared eight NoSQL databases and two promising New SQL databases based on the following criteria's: performance, reliability, integrity, security, query complexity, interoperability and cloud support.

Basing on this work, we choose the most efficient NoSQL system. It calculates the global score of each system, the comparative study start by calculating the weight of each criteria (performance, reliability, integrity, security and query complexity, interoperability and cloud support) by ROC (rank order centroid) method.

The result of ROC method is (0.37 for performance, 0.23 for integrity, 0.16 for reliability, 0.11 for interoperability, 0.07 for cloud support, 0.04 for query complexity, 0.02 for security). Then the authors calculates the global score, by attributing a note for each criteria for each NoSQL system.

The global score calculation formula is:

$$\begin{aligned}\mathbf{Globalscore} = (\mathbf{0.37}\,\text{performance}) + (\mathbf{0.23} * \text{integrity}) + (\mathbf{0.16} * \text{reliability}) \\ + (\mathbf{0.11} * \text{interoperability}) + (\mathbf{0.07} * \text{cloud support}) \\ + (\mathbf{0.04} * \text{query 4 complexity}) + (\mathbf{0.02} * \text{security})\end{aligned}$$

We focus only on the global score of the NoSQL systems, the results of the comparison are illustrated in the Fig. 1.

We focus only on the global score of the NoSQL systems, the results of the comparison of the global score shows that MongoDB system has obtained the best global score (4.96) [12].

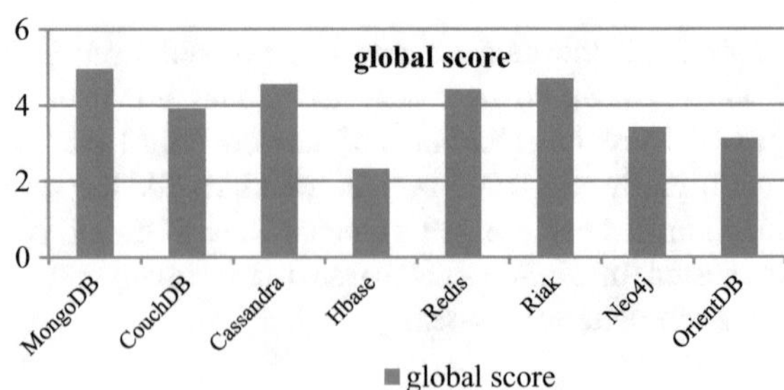

Fig. 1 The global score for the NoSQL systems

MongoDB, developed since 2007 by the 10gen software company, is a database management system that orients document written in C++ and very suitable for web applications, and MongoDB are the most popular NoSQL database [13].

In MongoDB, there are many techniques to achieve our model like indexing, aggregation. But the purpose in this paper is to present a model able to train and combining a real size of traffic of intrusion detection. So we decided to use a distributed programming technique MapReduce which is capable of managing large quantities of data [14, 15].

MapReduce is a programming model and this framework is an associated implementation for processing huge amounts of datasets.

MapReduce can deal with a large dataset by dividing in several tasks. It is constituted by two necessary functions, the first one (Map) is associated for emitting a key/value pair to generate a set of intermediate key/value pairs for each task of the dataset. And the second one (Reduce) merges all intermediate values, emitted from all tasks, related with the same intermediate key. The Reduce functions start only when all the map functions are finished.

In the next section, we present our model for training and combining semi-structured instances from the numerous sources based on the MapReduce framework in MongoDB.

4 Proposed Method

In order to train and combine the three datasets, our work is divided into two major steps:

A. The first one is to remove the redundancies instances and select the most pertinent features from triple datasets KDD99, DARPA99 and DARPA98.
B. The second is a vertical combination of triple instance files of the dataset (Fig. 2).

In our method, we presented a model so the size of those datasets does not reflect the real size of traffic of intrusion detection. Also, our model is able to manage heterogeneous format of data like son, picture. We describe our method in the following parts.

4.1 Pre-processing of the Dataset

Feature Selection

Intrusion detection system deals with huge amount of data which contains irrelevant and redundant features. Based on many works obtained, we have selected the most pertinent attributes and eliminate those that carry no information or information redundant.

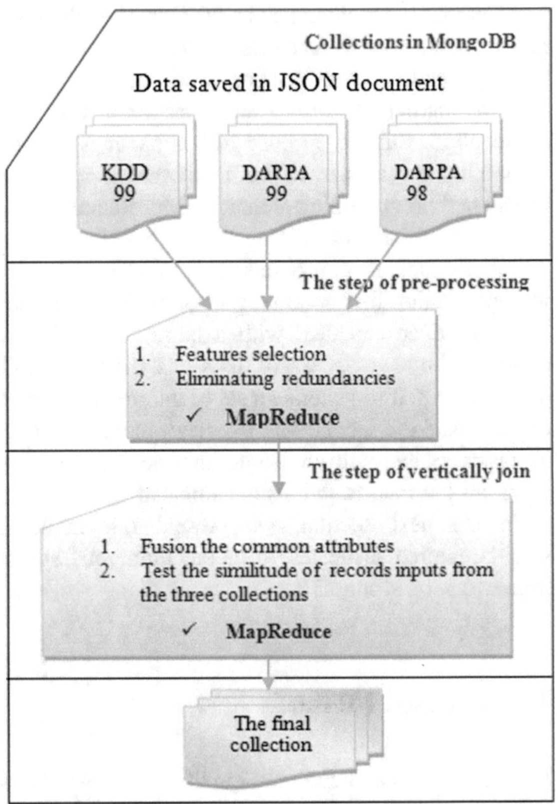

Fig. 2 Proposed method

We selected the most pertinent attributes in KDD that, according to [16, 17], we have used the method AFCM analysis (Factorial Multiple Correspondence Analysis) for this selection and choose those having the best gain. The attributes selected are: count, src_bytes, src_count, service, dst_host_same_src_port_rate, protocol_-type, dst_host_srv_count, dst_host_diff_srv_rate, dst_host_same_srv_rate.

In darpa 98, we selected only the useful features and eliminate information either redundant or not useful that identify the timestamp, source host and port, destination host and port [18].

Also, in darpa 99 and according to [19], the selection procedure of the features is managed by the ML algorithm. The attributes selected are: minfpktl, meanfpktl, maxfpktl, stdfpktl, minbpktl, protocol, fpacket.

The purpose of eliminating no-useful attributes is to attain a suitable classification for the evaluation of intrusion detection system and a better performance from the system.

Eliminating Redundancies

We should mention that in KDD 99, DARPA98 and DARPA 99, there are a large number of redundant recordings that causes a problem in terms of learning

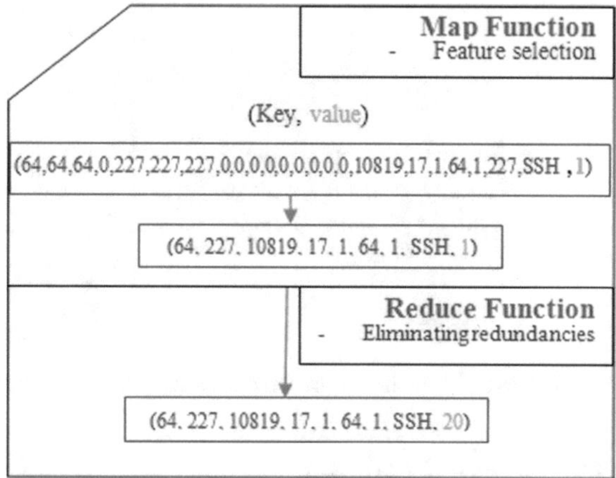

Fig. 3 The step of pre-processing with example from darpa99

algorithm and because of this redundancy, the learning algorithm biased toward the frequent records such as DOS and Probe attacks and consequently prevent them from learning the infrequent records such as U2R and R2L that are considered more noxious to the network.

In this step, the inputs of each data in MongoDB are stored in a different collection. Each line of the input data represents a connection and is saved as a document in the JSON format. And to achieve the step of pre-processing, we used a MapReduce under MongoDB for the two steps of pre-processing which we apply the procedure of eliminating redundancies just for the pertinent features (Fig. 3).

- **Map**: we applied a map function for each document in the collection. We emit for each document a pair:
  - **key** (the value of feature selection).
  - **value** (attribute 1 for each document to indicate that this document exists only once).
- **Reduce**: the Reduce function count the number of redundancies of each pair (key, value) emitted from function map by merging all intermediate values associated with the same intermediate key.
- The result is saved under the new collection.

So, in the map function, we selected the most pertinent attributes and in the reduce function, we eliminate the redundancy.

After we wrap up with the step of selecting the most pertinent attributes and removing the redundancies, we moved on to the next step to present a method with MapReduce in MongoDB to combine our datasets.

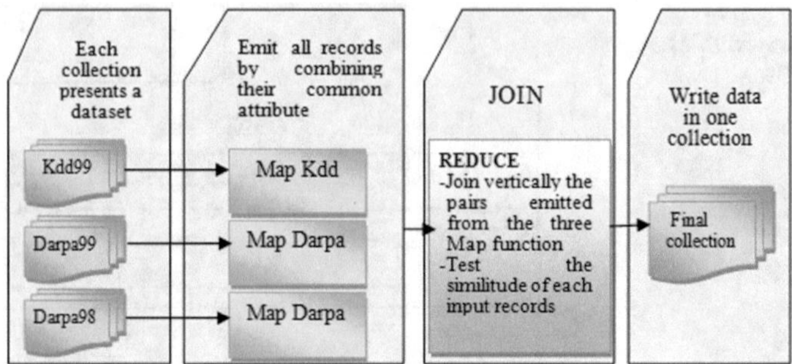

Fig. 4 Vertical combine of the three datasets in the same collection

4.2 Vertical Combine

Now, we have three datasets without redundancy. It just remains to merge the triple into a single dataset; the idea is described in the Fig. 4.

In this step our goal is to join vertically the three collections in MongoDB in one collection by fusion their common attributes. So the idea is to write a three map functions and only one reduce function.

- **Map Kdd, Map Darpa98 and Map Darpa99**: Each map function emit for each document:
  - key (name of each attribute by combining their common attribute).
  - value of each attribute (Fig. 5).
- **Reduce**: for each (key, value) sent by the three functions (map_kdd, map_-darpa98 and map_darpa99); A reduce function test the similitude of the values emitted from map functions by combining the similar records vertically for given more precision to each record (Fig. 6).

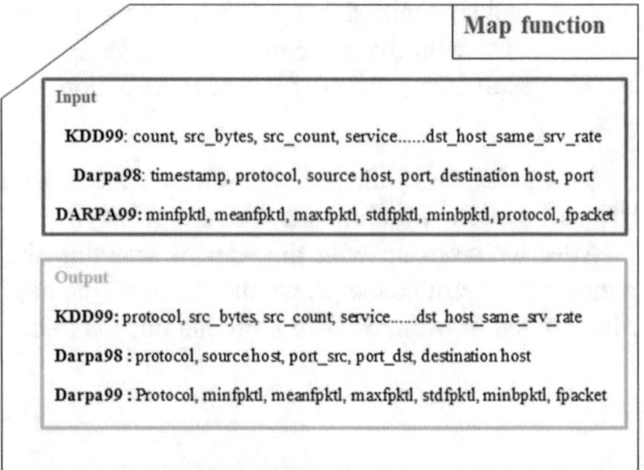

Fig. 5 The principle of combining the common attribute from the three datasets

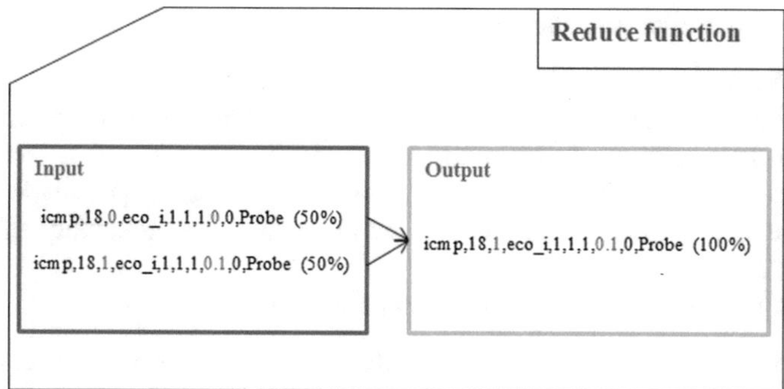

Fig. 6 The principle of the step of testing the similar records

- This result of all inputs records is saved vertically under the same new collection.

This fusion gives more diversity of attacks type, also gives more precision to each record by testing the similitude of each input record. All this improvement made necessary a big modification in intrusion detection rate. It increased the true positive rate and low down the false alarm rate.

Finally, when exporting the final collection, we used an aggregation to correct the problem of exporting a field from a subdocument.

5 Experiment and Results

In experimental result, we used the WEKA toolkit to analyse the dataset.

WEKA

Waikato Environment for Knowledge Analysis [20] is a popular suite of machine learning software written in Java, developed at the University of Waikato, New Zealand. This toolkit contains a collection of algorithms and visualization tools for data analysis and predictive modeling, with graphical user interfaces for easy access to these functions.

We are interested in a classify panel, which enables the application off classification and regression algorithms to the resulting dataset, to estimate the accuracy of the resulting predictive model (TP Rate), and to visualize erroneous predictions (FP Rate), receiver operating characteristic (ROC) curves, etc.

There are many algorithms to classify a dataset like Bayesian networks, naive Bayes, decision tree, etc.

For our experimentation we used the Bayesian Networks as K2 algorithm. We used this one to calculate the rate of intrusion detection (TP rate) and also the rate of false alarm (FP rate).

The K2 algorithm is described as follows:

K2 Algorithm
K2 learning algorithm is an algorithm with superior quality; it is specialized and useful for performance improvement in Bayesian network and a technique to optimize each node. The procedure of K2 algorithm is begins by a single node and its increment to connect others nodes for increasing the probability of network structure [21].

The following formula used to measure the performance:

- **TP Rate**: rate of true positives or detection rate (instances correctly classified as a given class)

$$TP = (\text{Total detected \_attacks/Total\_attacks}) * 100$$

- **FP Rate**: rate of false positives (instances falsely classified as a given class)

$$FP = (\text{Total\_misclassified\_process/Total\_normal\_attacks}) * 100$$

The deployment of Big Data technologies in the domain of intrusion detection is a new method. Recently, the work of Essid and Jemili [22] was published, their method serves to horizontally combine two sources of datasets (kdd99 and darpa 99) using the Big Data technique Map Reduce in Hadoop and they used the K2 algorithm, as a Bayesian network, for analysing their output.

In our work, we used the K2 algorithm, so to evaluate our work; we compared the results of our method with two methods using the same algorithm (K2):

- The method of Essid and Jemili [22].
- The method of Jemili et al. [23], they used a single dataset (kdd99) dataset.

When we compare the experimental results of Big Data technologies (MongoDB, Hadoop) with the work of Jemili et al. [23] in Table 3, we notice a remarkable amelioration of both MongoDB and Hadoop in term of detection rates.

For the results in Table 3 of MongoDB and the result of Hadoop, we can see that the two systems gives a good performance and we notice that MongoDB gives a

Table 3 Comparing the results of detection rates

Name of attack	Results with single dataset [23] (%)	Results with MongoDB (%)	Results of [22] Hadoop (%)
Normal	**87.68**	**99.80**	**97.40**
DOS	**88.64**	**98.40**	**99.96**
Probe	**99.15**	**99.90**	**97.02**
U2R	**6.66**	**88.90**	**93.32**
R2L	**20.88**	**98.10**	**97.41**
SSH	–	**65.30**	**57.1**
NOTSSH	–	**96.20**	**82.5**

Table 4 False positive in our proposed method

Types of attack	False positive (%)
DOS	0
Probe	0.2
U2R	0
R2L	0.2

certain amelioration better then Hadoop in term of results of detection rates in connections: Normal (our method gives 99.80% and Hadoop 97.40%), Probe, also for R2L (from 97.41 to 98.10%), SSH and NOTSSH. And it's necessary to indicate the number of connections for the both model Hadoop and MongoDB:

- For the output of Hadoop considered about 900,000 connections with size 120 MB.
- For the output of MongoDB considered about 600,000 connections with size 50 MB.

Although the difference between size for the both model, our model gives a better result then Hadoop and this improvement method is due to the step of selecting features when we selected the most pertinent features, the step of eliminate the redundancies from this pertinent features and the step of testing of similitude of connections emitted from the different sources, on the contrary Essid and Jemili [22] used all features for the two datasets.

For the two connections DOS and U2R, the method of Hadoop gives the results of detection rates better than our method with MongoDB. This improvement of method with Hadoop is due to the frequent number of connections of DOS and U2R.

Finally, we present the performance of our model in false positive results in the Table 4 and it's clear that our method gives high performance in terms of false positive.

So, the idea of using the Big Data storage as MongoDB can help in analyzing the traffic of intrusion detection, it would only take about few minutes to achieve the pre-processing of three datasets, also our model able to combining heterogeneous dataset like song or picture and other unstructured data. So the merging of different kind of datasets improves the results of performance indicators and subsequently the security system.

6 Conclusion and Future Work

The biggest challenge in the intrusion detection systems is the Big Data that are associated with large amounts of network traffic collected dynamically in the intrusion detection.

So, the big data techniques can necessary help the intrusion detection system for the management and the storage of several datasets and it is a new way for researchers to generate their own intrusion detection dataset.

In this paper, we used a NoSQL technology as MapReduce in MongoDB for the training and merging of different datasets kdd99, darpa1998 and darpa1999. Then, for the analyse of our dataset and the calculation of the performance metrics of our method, we used the K2 algorithm, as a Bayesian network, in order to compare it with other results: Our work showed a much better result than using a single dataset and better than Hadoop in some categories of attacks as Normal, Probe, R2L, SSH, NOTSSH. This compilation of big data techniques and intrusion detection system is greater and can necessary ameliorates the domain of security.

In future work, we will continue to develop our approach to merge others different distributed datasets by integrating MongoDB with others Big Data techniques and we will try to ameliorate the performance of detection rates and low false alarms. Also, we will integrate the MapReduce technique for developing distributed algorithms that would be able to process the new model of data.

References

1. Shanmugavadivu, R., Nagarajan, N.: Network intrusion detection system using fuzzy logic. Indian J. Comput. Sci. Eng. (IJCSE) **2**(1), 101–111 (2011)
2. Zekri, M., Meslati, L.S.: Immunological approach for intrusion detection. ARIMA J. **17**, 221–240 (2014)
3. Nadjaran Toosi, A., Kahani, M., Monsefi, R.: Network intrusion detection based on neuro-fuzzy classification. In: International Conference on Computing and Informatics, 2006. ICOCI '06, Kuala Lumpur, June 2006 (2006)
4. Hoque, M.S., Mukit, M.A., Bikas, A.N.: An implementation of intrusion detection system using genetic algorithm. Int. J. Netw. Secur. Appl. (IJNSA) **4**(2), 109–120 (2012)
5. Chickowski, E.: A case study in security big data analysis. http://www.darkreading.com/analytics/security-monitoring/a-case-study-in-security-big-data-analysis/d/d-id/1137299 (2012). Accessed Oct 2016
6. Richard, Z.T.M.Kh., Wald, R.: Intrusions detection and big heterogeneous data: a survey. J. Big Data 1, Article 115 (2015)
7. Azad, C., Jha, V.K.: Data mining in intrusion detection: a comparative study of methods, types and data sets. Int. J. Inf. Technol. Comput. Sci. (IJITCS) 75–90 (2013)
8. KDD Cup 1999 Data. http://kdd.ics.uci.edu/databases/kddcup99/. Accessed mars 2016
9. DARPA1998 Data. https://www.ll.mit.edu/ideval/data/1998data.html. Accessed mars 2016
10. DARPA1999 DATA. https://web.cs.dal.ca/~riyad/Site/Download.html. Accessed mars 2016
11. Lee, C.: An evaluation of machine learning techniques in intrusions detection. Doctoral Thesis in Computer Science, University of Vanderbilt, P6 (2007)
12. Dehbi, O.R., Talea, M., Batouta, Z.I.: An advanced comparative study of the most promising NoSQL and NewSQL databases with a multi-criteria analysis method. J. Theoret. Appl. Inf. Technol. **81**(3) (2015)
13. DB-Engines Ranking. http://db-engines.com/en/ranking. Accessed May 2016
14. MongoDB database. https://docs.mongodb.com/manual/core/map-reduce/. Accessed Sept 2016

15. Dean, J., Ghemawat, S.: MapReduce: simplified data processing on large clusters. In: OSDI'04: Sixth Symposium on Operating System Design and Implementation, San Francisco, CA, Dec 2004
16. Wei, W., Gombault, S., Guyet, T.: Towards fast detecting intrusions: using key attributes of network traffic. In: The Third International Conference on Internet Monitoring and Protection, Bucharest, vol. 13, pp. 86–91 (2008)
17. Al-Mamory, S.O., Jassim, F.S.: Evaluation of different data mining algorithms with KDD CUP 99 Data Set. J. Babylon. **21**, 2663–2681 (2013)
18. Zargar, G., Kabiri, P.: Identification of effective network features to detect Smurf attacks. In: Research and Development, SCOReD, p. 185 (2009)
19. Szabo, G.: Methods for efficient classification of network traffic. Thesis, Budapest University of Technology and Economics, p. 10 (2010)
20. Introduction to Weka. http://www.iasri.res.in/ebook/win_school_aa/notes/WEKA.pdf. Accessed 2016
21. Hernandez, J., Zarate, P., Dargam, F.: Decision Support Systems—Collaborative Models and Approaches in Real Environments, p. 61 (2011)
22. Jemili, F., Essid, M.: Combining intrusion detection datasets using MapReduce. In: The International Conference on Systems, Man, and Cybernetics (2016)
23. Jemili, F., Zaghdoud, M., Ahmed, M.B.: A framework for an adaptive intrusion detection system using Bayesian network. In: The IEEE International Conference on Intelligence and Security Informatics, USA, 2007 (2007)

Service Oriented Network Management with Knowledge-Based Network Management System in Fluctuating Environment

Yusuke Tanimura, Kazuto Sasai, Gen Kitagata and Tetsuo Kinoshita

Abstract The best advantage of cloud environment is elasticity. Although cloud-hosted ICT services gain an ability to flexibly change their size and granularity, deep knowledge and much experience are required for administrators to stably operate ICT services in such fluctuating environment. Various attempts on knowledge-based systems are made to reduce burden on administrators. However, knowledge-based systems necessitate management knowledge appropriate to ICT service configuration, and consequently a burden on maintaining knowledge base is not negligible in fluctuating environment. We propose service oriented network management with knowledge-based network management support system. Proposed system is modularized and can be applied to fluctuating environment with less burden. We demonstrate the effectiveness of our proposal through experiments using implemented prototype system.

Keywords Network and systems management · Multi-agent system · Knowledge-based system · Active information resource

---

Y. Tanimura (✉)
Graduate School of Information Sciences, Tohoku University, Sendai, Japan
e-mail: tanimura@k.riec.tohoku.ac.jp

K. Sasai · G. Kitagata · T. Kinoshita
Research Institute of Electrical Communication, Tohoku University, Sendai, Japan
e-mail: kazuto@riec.tohoku.ac.jp

G. Kitagata
e-mail: minatsu@riec.tohoku.ac.jp

T. Kinoshita
e-mail: kino@riec.tohoku.ac.jp

R. Lee (ed.), *Software Engineering, Artificial Intelligence, Networking and Parallel/Distributed Computing*, Studies in Computational Intelligence 721,
DOI 10.1007/978-3-319-62048-0_3

1 Introduction

The long-established way for constructing ICT services is to install an operating system directly on a computing resource. Although this is simple and intuitive to be managed, ICT services constructed based on the way have a limitation on flexibility. In recent years, a methodology for constructing ICT services based on virtualization technology have been widespread. By deploying virtual machines on a hypervisor, administrators of the network system can flexibly construct and reconfigure ICT services. A concept of cloud computing is emerging and becoming popular with such background. A key feature of cloud computing is its elasticity. Cloud-hosted ICT services constructed based on virtualized computing resource have ability to adapt flexibly to fluctuated demands of users.

In cloud computing, we consider that there are two types of elasticity: *vertical elasticity* and *horizontal elasticity*.

Vertical elasticity is ability to scale an ICT service according to the magnitude of user demand. Administrators can scale cloud-hosted ICT services by configuring a size of virtual machine, or the number of replicated virtual machines through a load balancer. Most hypervisors have functions for automatically scaling ICT services. Moreover, various research and development activities are conducted to realize effective ICT service scaling [3, 6].

Horizontal elasticity is ability to rapidly build or remove ICT services for responding to diverse needs of users. Because of this characteristic, the number of ICT services which have to be managed fluctuates frequently in cloud environment. In order to follow up the fluctuating user demand, various automation tools for configuration management are researched and developed [12]. These tools make it easier for administrators to build ICT services quickly.

While ICT services gain elasticity by automation tools for configuration management, there still remains heavy burden on administrators, especially in environment that makes the most of horizontal elasticity. Network management tasks are not only configuration management. In environment which based on horizontal elasticity, various ICT services are automatically built and removed one after another in a short period of time. Deep knowledge and much experience are required for administrators to stably operate ICT services in such fluctuating environment. Thus, an intelligent support system for administrators is vital to utilize benefit of elasticity in cloud environment. This is the focus of this paper.

In this paper, we propose service oriented network management with knowledge-based management support system. We introduce a knowledge-based network management system (KNMS) that has human expertise in its knowledge base and has ability to behave like an expert administrator. In order to support administrators of fluctuating network system, we modularize the KNMS and make it easier to rearrange knowledge base responding to additions and deletions of ICT services.

2 Related Work

Various research and development are conducted on a network management system (NMS) as a tool to reduce the burden on network and system administrators. NMSs commonly used in actual network management are designed with a focus on the function of monitoring the state of network devices using monitoring protocol. Such NMSs can detect abnormal changes of management targets and report to administrators by setting criteria for defining normal state. However, tasks such as identifying detailed fault causes and thinking out countermeasures need to be carried out by administrators themselves.

From this perspective, various attempts are made to enhance the ability of the NMS and enlarge the scope of automation by applying knowledge derived by expert administrators. In this paper, we call such NMSs Knowledge-based Network Management Systems (KNMSs). The KNMS has knowledge base and accumulate knowledge described as rules or policies. KNMSs work autonomously by utilizing human expertise and alleviate a burden of network management tasks.

Autonomic Network Management System (ANMS) [4] is a popular approach to implementation of KNMS. ANMS is a concept based on the idea of Autonomic Computing [1], inspired by the autonomic nervous system. NMSs based on the concept of ANMS have ability to keep the normal state of managed system. A key point of the concept of ANMS is a control loop called MAPE-K (Monitor-Analyze-Plan-Execute over a Knowledge) feedback loop. To realize autonomic features, ANMSs utilize knowledge derived from administrators for control loop. In addition, various approaches based on artificial intelligence are conducted to apply human knowledge to network operation [5, 7, 9, 10].

However, to effectively use a KNMS, administrators are required to maintain knowledge base according to the change in configuration of the managed system. Since knowledge unsuitable for system configuration causes malfunction and degradation in system performance, administrators must pay close attention to consistency of knowledge. Conventional attempts to substantiate a concept of KNMSs are intended to be applied to static environments whose configuration do not frequently change, hence the burden of maintaining knowledge base is not regarded as important. As mentioned above, when operating cloud-hosted ICT services, the system configuration frequently changes to respond to fluctuating needs of users. Therefore, in operation of cloud-hosted ICT services with support by KNMSs, the burden on administrators involved with maintaining knowledge base is not negligible.

3 Proposal

We propose a KNMS which can be applied to fluctuating environment. We reinforce a concept of Active Information Resource based Network Management System (AIR-NMS) [7], and make it easy for administrators to rearrange its knowledge base. First, we describe the concept of AIR-NMS.

3.1 AIR-NMS

An Active Information Resource (AIR) is a distributed information resource that is enhanced with a Knowledge of Utilization Support (KUS) and a Function of Utilization Support (FUS), which act as an autonomous agent [2]. The KUS consists of meta-level knowledge, i.e., knowledge for handling information resources and cooperation knowledge with other AIRs, and the FUS consists of various functions to process its information and to communicate with other AIRs. AIRs can cooperate with each other and process complex information actively and autonomously with these utilization supports. By using the concept of AIR, it is possible to deputize tasks necessary for utilization of information resources, thus the burden on users can be reduced.

Figure 1 shows a conceptual model of AIR-NMS. The AIR-NMS is a network management system based on the concept of AIR [7]. AIR-NMS consists of Information AIRs (I-AIRs) and network management Knowledge AIRs (K-AIRs). An I-AIR manages the network status information acquired from network equipment as its information resource, such as IP addresses, application settings, server logs, etc. A K-AIR manages the network management knowledge which human administrators have learned through their experience. In the concept of AIR-NMS, network information (I-AIR) and management knowledge (K-AIR) cooperate with each other and support management tasks to reduce the burden on administrators. As an practical application of AIR-NMS, studies focusing on support for fault management are conducted [2, 8].

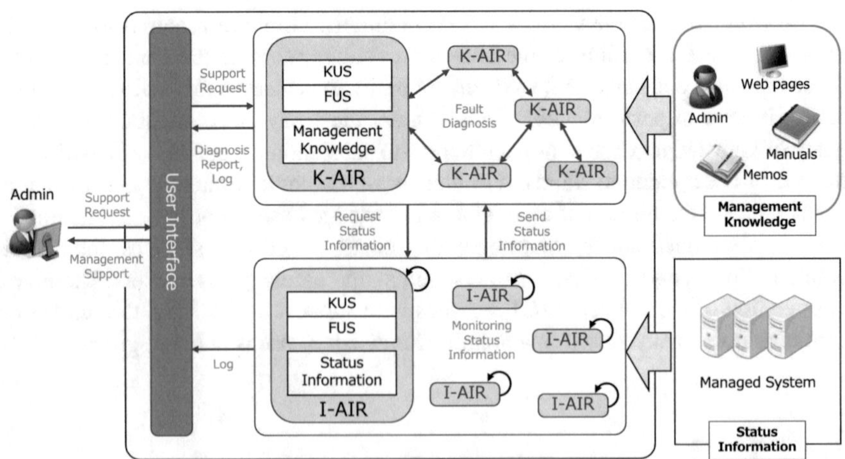

Fig. 1 Conceptual model of AIR-NMS applied to fault management tasks

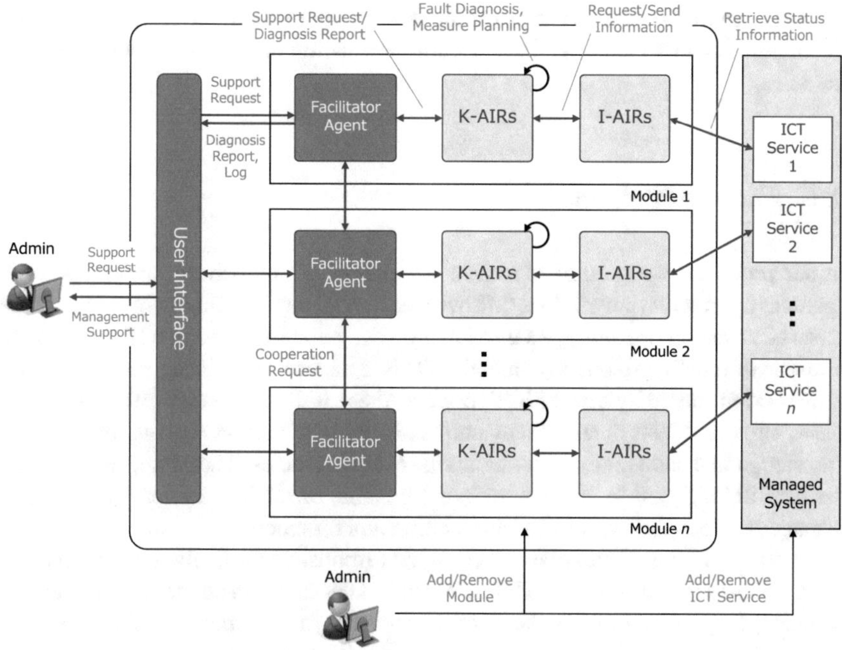

Fig. 2 Conceptual model of AIR-NMS improved for cloud-hosted ICT services operation

3.2 *Service Oriented Network Management with AIR-NMS*

Figure 2 shows a conceptual model of our proposed system. In this proposal, we modularize the AIR-NMS in order to flexibly add or revise knowledge according to changes in configuration of cloud-hosted ICT services, which is unlike conventional KNMSs based on monolithic architecture. We consider that changes in system configuration happen when a new demand for ICT services occurs. Thus, we model a module for each ICT service. A module contains K-AIRs, I-AIRs and a Facilitator-Agent. When a change in system configuration occurs, the AIR-NMS follows the change by connecting/disconnecting module. When an ICT service is newly added or becomes unnecessary, the AIR-NMS follows the change in system configuration by connecting or disconnecting the module related to the ICT service.

In the proposed system, K-AIRs are divided and managed for each ICT service, and the scope to consider the consistency of knowledge is limited within each module. Thus, it becomes possible to handle knowledge solely focused on an ICT service which is newly added or becomes unnecessary, independent of all other ICT services. Therefore, the burden due to knowledge management is reduced, and administrators can add or revise knowledge easily to AIR-NMS in the operation of fluctuating environment whose system configuration frequently changes.

Furthermore, in order to improve the operability of the module, we introduce a Facilitator-Agent which has a function for verification of AIR cooperation between modules.

3.3 Facilitator-Agent

In our proposal, we attempt to reduce administrator's burden on knowledge management by modularizing AIR-NMS with each ICT service. On the other hand, in a scene of an actual operation, a same function can be shared among ICT services. An example of function frequently shared is DNS, database, etc. When a problem occurs in a function shared by several ICT services, these ICT services are affected. In such cases, sufficient fault management support is difficult without sharing management knowledge and status information among several modules. Therefore, we introduce the Facilitator-Agent to achieve both, the benefits of modularization, and the fault management support with AIRs cooperation among modules.

As shown in Fig. 2, Facilitator-Agents are responsible for facilitating and mediating AIR cooperation among modules. When it is necessary to cooperate AIRs among modules, they cooperate via the Facilitator-Agent in each module. To verify AIR cooperation, the Facilitator-Agent have following functions.

- Function to make a list of AIRs in a module: A Facilitator-Agent makes a list of K-AIRs and I-AIRs in a module, namely, a list of meta-information about knowledge and status information holding in the module.
- Function to examine AIR cooperation among modules: A Facilitator-Agent examine whether a cooperation request from AIRs of other modules can be processed in own module. The examination is executed based on the AIR list, and only when judged as acceptable, the cooperation request is forwarded to AIRs in own module.

In the previous system, since all AIRs work in a same place, cooperation requests are broadcasted to all AIRs. On the other hand, in the proposed system, AIRs are modularized and divided for each ICT service module. Moreover, cooperation requests are transferred by Facilitator-Agent only to a module which can accept requests. Therefore, the number of cooperation requests can be minimized in the proposed system. This means that the burden on AIR management including knowledge management can be reduced.

4 Design of Prototype System

We design a prototype system to evaluate our proposal. The prototype system is designed to implement on a repository-based multi-agent framework ADIPS/DASH [11].

4.1 K-AIR and I-AIR

A K-AIR handles management knowledge about troubleshooting obtained from expert administrators or management manuals. The K-AIR has KUS for cooperating with other AIRs which have related management knowledge or status information required for fault management, and has FUS for extracting and parsing management knowledge described in a text file.

We design three types of K-AIRs as with previous work [7], Ksc-AIR, Kcd-AIR and Kcm-AIR. Figures 3, 4 and 5 show the description examples of management knowledge. Each type of K-AIR handles the following management knowledge:

```
<sc symptom="SIP client cannot make calls">
   <info>SIP-client_ID</info>
   <cause>SIP server process down</cause>
   <cause>SIP connection port not open</cause>
   <cause>Client ID not permitted to call</cause>
</sc>
```

Fig. 3 Description example of Ksc

```
<cd cause="SIP connection port not open">
  <dm>
    <p>get #SIP-port_num# cmd #grep ^port /etc/asterisk/sip.conf
    |awk -F "=" '{print $2}'# login #SIP-server_IP#</p>
    <p>get #SIP-port_state# cmd #grep val(SIP-port_num)
    /etc/sysconfig/iptables |awk -F " " '{print $NF}'#
    login #SIP-server_IP#</p>
    <p>true (#SIP-port_state# -ne ACCEPT)</p>
  </dm>
  <dr>
    Port number #SIP-port_num# is not open at server #SIP-server_IP#.
  </dr>
</cd>
```

Fig. 4 Description example of Kcd

```
<cm cause="SIP connection port not open">
  <m>
  Open UDP port #"^port="value"$"#.
  1. Login server #SIP-server_IP# as superuser.
  2. Open /etc/sysconfig/iptables and add a discription of below.
   -A INPUT -m state --state NEW -m udp -p udp --dport #"^port="value"$"#
  3. Reload iptables with "/etc/init.d/iptables restart".
  </m>
</cm>
```

Fig. 5 Description example of Kcm

- Ksc (symptom-cause): Cause assuming—assumes the conceivable causes from observed symptoms or detected faults.
- Kcd (cause-diagnose): Cause diagnosing—diagnoses the exact causes of the faults and presents the diagnosis reports.
- Kcm (cause-means): Measure planning—plans the countermeasures against the identified causes and presents them.

When an administrator ask the AIR-NMS to support troubleshooting, K-AIRs start the diagnosis. First, Ksc-AIRs derive assumed failure causes and send messages to Kcd-AIRs where the assumed causes can be diagnosed. Then Kcd-AIRs verify the assumed causes and identify a root cause. If Kcd-AIRs identify any cause, they send messages to Kcm-AIRs to ask for a countermeasure. Finally Kcm-AIRs generate the countermeasure and present to the administrator. By executing the presented countermeasure, the administrator can recover the fault.

An I-AIR has KUS for monitoring managed object and cooperating with other AIRs, and has FUS for retrieving, processing, and storing status information of managed object. While K-AIRs are in process of troubleshooting support, K-AIRs inquire the network status information of the I-AIRs, which is necessary for the diagnosis. In this case, I-AIRs retrieve the required information from the managed system and reply to the K-AIRs.

4.2 Facilitator-Agent

We design a protocol, for a function to make a list of AIRs in a module, to inquire of AIRs in same module possessing meta-information about knowledge and status information.

To search Facilitator-Agent of other modules, Facilitator-Agents use a name server function of ADIPS/DASH. The Facilitator-Agent periodically look up other module and adapts to changes of configuration of ICT services. Figure 6 shows an example of meta-information that can be acquired by this protocol, and meta-information is associated with the AIR name of the information source. A cooperation request from Facilitator-Agent of other module is verified based on this list.

```
(SIP server process down, Ksc-AIR.201412111553015:w2:AIR-NMS-PC-1)
(SIP connection port not open, Kcd-AIR.201412111553020:w3:AIR-NMS-PC-1)
(Client ID not permitted to call, Kcd-AIR.201412111553023:w3:AIR-NMS-PC-1)
```

Fig. 6 Example of meta-information associated with AIR name

5 Experiments

We implemented a prototype system based on the design mentioned above. We also implemented a system based on a previous work [7] for evaluation with a comparison of experimental results. We conducted experiments about troubleshooting tasks with two types of AIR-NMS, *proposed system* and *previous system*.

AIRs and agents in an AIR-NMS implemented based on the ADIPS/DASH framework cooperate each other by exchanging of messages. Thus, we compared an efficiency of AIR/agent cooperation, by measuring the number of messages generated during troubleshooting support. In addition, we simulated an addition of knowledge to AIR-NMS, and evaluate the burden on knowledge management.

Figure 7 shows the construction of managed system. In managed system, five types of ICT services were in operation. Each ICT service consists of one or more virtual machines. In this experimental environment, the Web service depended on the DB service, and the DB service and the Mail service depended on the Storage service.

Table 1 shows the number of management knowledge prepared for the experiment. In the proposed system, management knowledge was modularized for each ICT service. On the other hand, in the previous system, all management knowledge was gathered at a single knowledge base.

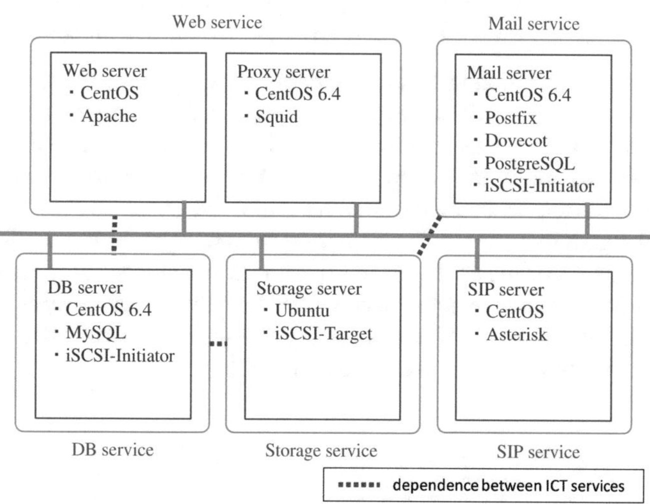

Fig. 7 Experimental environment where ICT services are managed

Table 1 The number of management knowledge for each ICT service

	# of management knowledge		
	Ksc	Kcd	Kcm
Web service	2	11	11
DB service	1	5	5
Storage service	1	4	4
SIP service	1	6	6
Mail service	3	13	13
Total	8	39	39

5.1 The Number of AIR/agent Messages During Troubleshooting Support

We conducted experiments on troubleshooting for following five types of failure cause:

- DB server process down (failure symptom: Web page cannot be browsed).
- HTTP server process down (failure symptom: Web page cannot be browsed).
- Proxy server process down (failure symptom: Web page cannot be browsed).
- SIP server process down (failure symptom: SIP client cannot make calls).
- POP/IMAP server process down (failure symptom: Email cannot be received).

Figure 8 shows one of the result of troubleshooting support presented by the proposed system. For every failure case, the proposed system and the previous system presented one or more countermeasure. From this result, we confirmed that the proposed system had ability to support administrators as well as the previous system, by utilizing AIRs among modules through Facilitator-agents.

Figure 9 shows that the number of AIR/agent messages during troubleshooting support in each experiments. In every experiment, the number of messages in the proposed system was greatly reduced compared with that in the previous system. We considered that this is because the proposed system was modularized and a broadcast domain of cooperation request was restricted in each module.

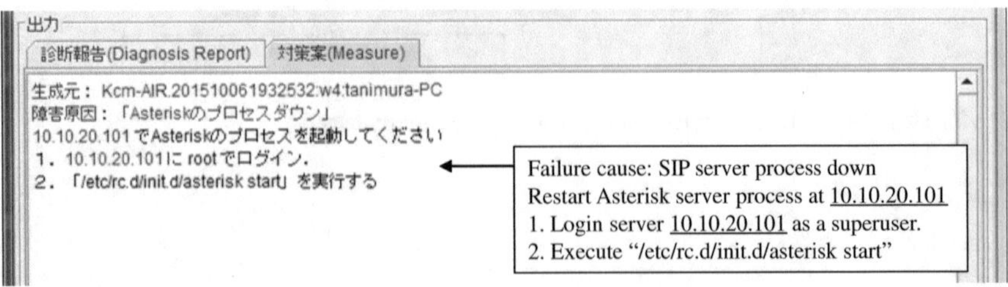

Fig. 8 Output countermeasure (proposed system)

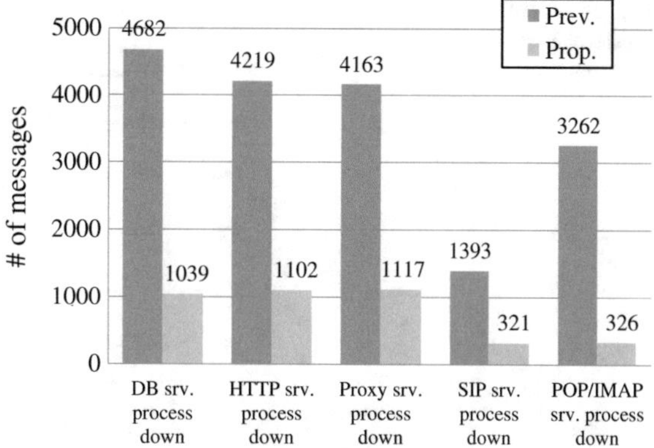

Fig. 9 The number of AIR/agent message during troubleshooting support

The AIR-NMS is a multi-agent based KNMS, hence the management of knowledge consistency in the AIR-NMS is equivalent to directing messages between AIRs/agents. Therefore, in the proposed system, the burden on knowledge management is reduced because the number of messages which must be controlled by administrators are less than that of the previous system.

5.2 *Simulation of Knowledge Base Management*

To compare the burden on knowledge management, which occurred when ICT services were added to managed network, we conducted simulation experiments. In this simulation, we expressed the amount of the burden on administrators as the number of knowledge that must be confirmed thier consistency.

When a kth ($k = 1, 2, \ldots$) ICT service is added to the managed network, the burden on administrators who use the previous system N_k, and the burden on administrators who use the proposed system N'_k are:

$$N_k = \sum_{i=1}^{k} n_{\mathrm{sc}_i} + \sum_{i=1}^{k} n_{\mathrm{cd}_i} + \sum_{i=1}^{k} n_{\mathrm{cm}_i} \tag{1}$$

$$N'_k = \sum_{i=1}^{k} n_{\mathrm{sc}_i} + n_{\mathrm{cd}_k} + n_{\mathrm{cm}_k} \tag{2}$$

where n_{sc_i}, n_{cd_i} and n_{cm_i} denote the number of Ksc, Kcd and Kcm for the ith ($i = 1, 2, \ldots, k$) ICT service.

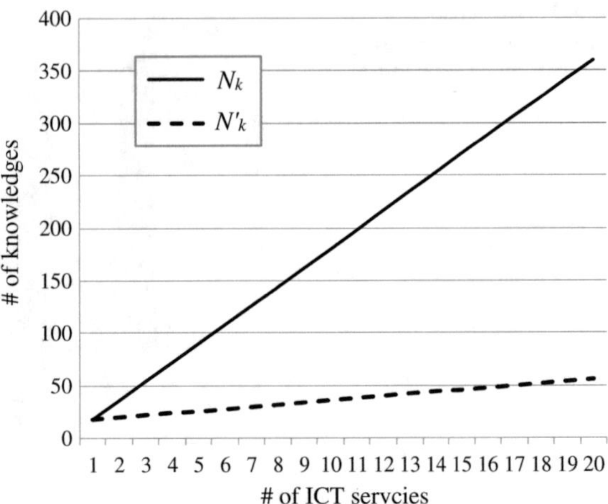

Fig. 10 Comparison of burden on knowledge management

In the previous system, when new knowledge is added to the knowledge base, administrators must confirm the consistency of all Ksc, Kcd and Kcm. Thus the burden on administrators increases accumulatively. On the other hand, in the proposed system, although administrators must confirm the consistency of Ksc between all modules considering the sharing of functions between ICT services, the confirmation task for Kcd and Kcm is limited within the newly added *k*th module.

Figure 10 shows the burden on administrators, which is estimated based on the experiments using the implemented system mentioned above. We set values of n_{sc_i}, n_{cd_i} and n_{cm_i}, from the average number of knowledge used in the experiment (see Table 1), namely, $n_{\mathrm{sc}_i} = 2$, $n_{\mathrm{cd}_i} = 8$ and $n_{\mathrm{cm}_i} = 8$. Then, Eqs. 1, 2 are rewritten as follows:

$$N_k = 18k \tag{3}$$

$$N'_k = 2k + 16 \tag{4}$$

Figure 10 shows that the proposed system, the burden on administrators was greatly reduced compared with the previous system, even when the number of ICT services was increased.

5.3 Discussion

From experimental results, it was shown that the burden on administrators accompanied by knowledge management of a KNMS, was reduced in the proposed system because of modularization. Consequently, we considered that our proposal solved the problem of the KNMS, which is about knowledge management in fluctuated system configuration, and our proposal can be applied to fluctuating environment.

6 Conclusion

In this paper, we proposed service oriented network management with knowledge-based network management support system. We reinforced the concept of AIR-NMS with modular architecture, to improve the manageability of knowledge base. Experiments using implemented prototype system were conducted and experimental results indicated that the burden due to knowledge management was reduced in proposed system. Thus it was confirmed that the proposed system can be applied to fluctuating environment and support administrators effectively. In near future, we plan to enhance an ability of AIRs by introducing expert's heuristics for investigating a configuration of ICT service.

Acknowledgements This work was supported by Council for Science, Technology and Innovation (CSTI), Cross-ministerial Strategic Innovation Promotion Program (SIP), "Enhancement of societal resiliency against natural disasters" (Funding agency:JST).

References

1. Kephart, J.O., Chess, D.M.: The vision of autonomic computing. Computer **36**(1), 41–50 (2003)
2. Kinoshita, T.: Agent-based active information resource and its applications. In: Databases in Networked Information Systems, pp. 143–156. Springer (2010)
3. Lorido-Botran, T., Miguel-Alonso, J., Lozano, J.A.: A review of auto-scaling techniques for elastic applications in cloud environments. J. Grid Comput. **12**(4), 559–592 (2014)
4. Movahedi, Z., Ayari, M., Langar, R., Pujolle, G.: A survey of autonomic network architectures and evaluation criteria. IEEE Commun. Surv. Tutor. **14**(2), 464–490 (2012)
5. Ross, M., Covo, A., Jr Hart, C.: An Al-based network management system. In: Seventh Annual International Phoenix Conference on Computers and Communications, 1988 Conference Proceedings, pp. 458–461. IEEE (1988)
6. Salah, K., Elbadawi, K., Boutaba, R.: An analytical model for estimating cloud resources of elastic services. J. Netw. Syst. Manag. **24**(2), 285–308 (2016)
7. Sasai, K., Sveholm, J., Kitagata, G., Kinoshita, T.: A practical design and implementation of active information resource based network management system. Int. J. Energy Inf. Commun. **2**(4), 67–86 (2011)
8. Tanimura, Y., Sveholm, J., Sasai, K., Kitagata, G., Kinoshita, T.: A knowledge-based support method for autonomous service operations after disasters. In: 2013 IEEE/ACIS 12th International Conference on Computer and Information Science (ICIS), pp. 229–233. IEEE (2013)
9. Tesauro, G., Chess, D.M., Walsh, W.E., Das, R., Segal, A., Whalley, I., Kephart, J.O., White, S.R.: A multi-agent systems approach to autonomic computing. In: Proceedings of the Third International Joint Conference on Autonomous Agents and Multiagent Systems - AAMAS'04, vol. 1, pp. 464–471. IEEE Computer Society (2004)

10. Tran, H.M., Schönwälder, J.: Distributed case-based reasoning for fault management. In: IFIP International Conference on Autonomous Infrastructure, Management and Security, pp. 200–203. Springer (2007)
11. Uchiya, T., Hara, H., Sugawara, K., Kinoshita, T.: Repository-based multiagent framework for developing agent systems. In: Transdisciplinary Advancements in Cognitive Mechanisms and Human Information Processing, pp. 60–79 (2011)
12. Wettinger, J., Behrendt, M., Binz, T., Breitenbücher, U., Breiter, G., Leymann, F., Moser, S., Schwertle, I., Spatzier, T., et al.: Integrating configuration management with model-driven cloud management based on TOSCA. In: Closer, pp. 437–446 (2013)

A Topic Structuration Method on Time Series for a Meeting from Text Data

Ryotaro Okada, Takafumi Nakanishi, Yuichi Tanaka, Yutaka Ogasawara and Kazuhiro Ohashi

Abstract In this paper, we present a dialogue structure analysis method to visualize the transition of topics in a meeting as the one of dialogue process representation. Our method extracts topics in a meeting on time series. In addition, we define an index to assess the importance of the whole meeting in each phase. By this index, we can represent important phases in the meeting. In organizations such as companies, it is important to improve the efficiency of a meeting, because the meeting time occupies a large proportion in business hours. We should analyze contents and flows of remarks in dialogue on meetings in order to improve efficiency of a meeting. Generally, improving the efficiency of a meeting is improving the form of a meeting, such as pre-sharing of documents, keeping time, clarification of roles of members, and appointing a facilitator. Our method provides the one of the visualization for the flow of remarks in dialogue in a meeting. In this paper, we also represent some preliminary experiment by using text data for actual meetings.

Keywords Topic structuration · Efficiency of a meeting · Dialogue · Transition of topics · Time-series data

R. Okada · T. Nakanishi (✉)
Center for Global Communications (GLOCOM) International University of Japan, Tokyo, Japan
e-mail: takafumi@glocom.ac.jp

R. Okada
e-mail: rokada@glocom.ac.jp

Y. Tanaka · Y. Ogasawara · K. Ohashi
ITOKI Corporation, Tokyo, Japan
e-mail: tanaka4uu9@itoki.jp

Y. Ogasawara
e-mail: ogasawara2hd9@itoki.jp

K. Ohashi
e-mail: ohashi23g6@itoki.jp

R. Lee (ed.), *Software Engineering, Artificial Intelligence, Networking and Parallel/Distributed Computing*, Studies in Computational Intelligence 721,
DOI 10.1007/978-3-319-62048-0_4

1 Introduction

In day-to-day production activities, dialogues play very important roles in order to exchange information and create new ideas. In organizations such as companies, many people have some meeting for dialogue to each other. The meeting time occupies a large proportion in business hours. According to the survey by NTT Data Institute of Management [1], in the case of Japanese companies, the time they spend on meetings occupies 15.4% on average in business hours. In addition, the survey asked them to point out problems of meetings. The important problems they mentioned are "There are often unnecessary meetings", "There are too long meetings", and "There are too frequently meetings." We infer from these facts that efficiency improvement of meetings is not enough in spite of meetings occupy much of the working hours.

Generally, improving the efficiency of a meeting is improving the form of a meeting, such as pre-sharing of documents, keeping time, clarification of roles of members, and appointing a facilitator. We should analyze contents and flows of remarks in dialogue on meetings in order to improve efficiency of a meeting.

In the method of analysis for a meeting, the ethnographic methods [2] have been mainly studied. In these methods, researchers continue to observe the behavior of attendances during the meeting. The methods require the researcher to attend the meeting actually for observation. Generally speaking, it is difficult to keep analyzing by using these methods because of high analysis cost for observing the meeting. Moreover, since the methods are qualitative studies based on observation by human, we cannot practically implement as computer programs at this stage.

On the other hand, it becomes easier to obtain voice data by improvement and cost reduction of the microphone. In addition, it is easy to obtain text data from voice data by improvement of voice recognition technology recently. Based on this background, we design new indexes for efficiency in a meeting by data mining techniques for these text data. An efficiency of a meeting means the elimination of useless meeting, shortening time of meeting, and the promotion of more productive and meaningful meeting.

Facilitating and reviewing a meeting are important for the efficiency of a meeting. It is important to objectively know process of arriving at decision for dialogue by facilitating and reviewing a meeting.

In this paper, we present a dialogue structure analysis method to visualize the transition of topics in a meeting as the one of dialogue process representation. Our method extracts topics in a meeting on time series. In addition, we define an index to assess the importance of the whole meeting in each phase. By this index, we can represent important phases in the meeting. We also present some preliminary experiment by using text data for actual meetings.

The feature of our method is an automatically extracting and a structuring of topics for meetings by focus on the flow of remarks in dialogue from a text data generated by voice data in the meeting.

This paper is organized as follows: In Sect. 2, we introduce related works. In Sect. 3, we present our structuration method of meeting focused on time series. In Sect. 4, we present some preliminary experiments. In Sect. 5, we reach a conclusion.

2 Related Work

There are many factors for analyzing a meeting such as contents they talked, volume of voice, frequency characteristic of voice, attributes of members, behavior of speakers and listeners, etc.

The ethnographic methods [2] consider these factors to evaluate a quality of a meeting. These methods are based on comprehensive observation in dialog analysis. These methods are qualitative evaluation methods and largely depend on the experiences of an observer.

There are also quantitative evaluation methods in dialogue analysis. Researches on analytical methods focusing on non-verbal information such as document (for example, [3–5]) are actively conducted. These methods considers non-verbal factors such as the timing of utterance and silence, intonation, pitch/size of voice, speed of speak, etc. to evaluate features of dialogue. These non-verbal information can be easily extracted automatically from voice data.

Our method is the one of analysis methods focusing on language information. In recent years, the performance of speech recognition has much increased due to the development of machine learning technology. It is predicted that various dialogues is accumulated as text data in the near future. When text mining is applied to these text data, it becomes possible to look back on the dialog effectively and analyze meaning.

One of the best-known methods of extracting topics in text analysis is LDA (latent dirichlet allocation) [6]. LDA is the one of Topic model which is a model to estimate topics using stochastic method. Topics model estimates the latent topics that span multiple documents. In our method, we divide an entire text to segments and extract topics that span multiple segments. The point of extracting topics from multiple parts is common to our method and Topic model. However, our method aims not only to extract topics, but also to review when a topic was spoken, because we focus on improving a meeting process. As studies to divide whole text to parts by topic, Text segmentation [7–10] methods are usable. Text segmentation methods first divide the text into sentences. Then, the methods measure similarities among these sentences. Finally, the methods identify where the topic switched. As methods of text segmentation, TextTiling [7], C99 [8] and etc. are well known. Furthermore, in order to solve the problem of sparseness of word vectors in these methods, studies are also being conducted to compress dimensions using a topic model. There are cases where topic models are applied to TextTiling [9] and applied to C99 [10]. The research of text segmentation is in common with our research in that it estimates what and where is spoken in an entire text.

However, in our research, we do not assume that a segment does not necessarily have a common topic with other segments. This is a point of difference from the research of text segmentation. In existing researches on text segmentation have dealt with texts written by human. In contrast, in our research, we target text transcribed from a speech spoken at a meeting. Sometimes, we also have a meaningless conversation in a meeting. A creative process necessarily includes trial and error. Since we focus on the process of a meeting in our research, we aim to analyze not only important conversation but also unimportant conversation.

The purpose of the researches of text segmentation is basically to divide the text, and it is often used as preprocessing of content analysis. It is possible to apply the text segmentation methods as preprocessing of our method.

3 A Topic Structuration Method for a Meeting

In this section, we present our method, a structuration method of meeting focused on time series. Our method extracts topics in a meeting on time series. In addition, we define an index to assess the importance of the whole meeting in each phase. By this index, we can represent important phases in the meeting.

First, we give an overview of our method in Sect. 3.1. Next, we show the design of a mathematical formulation in 3.2. Finally, we introduce three output formats method for representing the flow of the conversation as analysis results derived by using our method.

3.1 Overview

The overview of our proposed system is shown in Fig. 1. In our system, an input is text data transcribed from speech. First, our system splits the text data into phases in any number n and converts each phase into each vector as a common process. Each phase represents in each vector. Each element of the vector is a TF-iPF weighted value of each word shown in (3.2.3).

The system finally provides three type of visualization based on structure of a meeting as follows: (1) Phase feature words, (2) Meeting-outline, and (3) Importance graph.

Phase feature words consist of feature words to which an arbitrary phase belongs and topic words of the group to which the phase belongs. By the Phase feature words, we can find important keywords and topics of each phase in the meeting. The Phase feature words are obtained from the group to which the phase belongs, the topic words in the group, and the feature words in the phase.

Meeting-outline consists of the correspondence relation between each group and each phase and the list of topic words of each group in the whole meeting. By the Meeting-outline, we can find the flow of conversation in the meeting. The

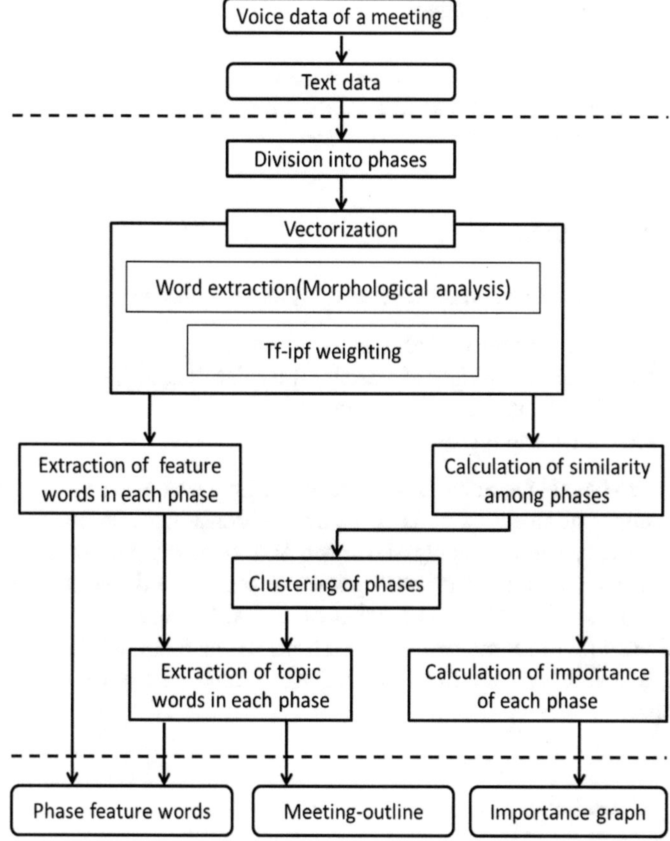

Fig. 1 Overview of our system. The input of the system is a voice data of a whole meeting. The system finally outputs three visualizations, "Phase feature words", "Meeting-outline" and "Importance graph"

Meeting-outline is obtained from the group to which the phase belongs, and the topic words in that group.

Importance graph consists of an importance of each phase in the whole meeting. By the Importance graph, we can find important scenes (that is, phases) in the meeting. The Importance graph is obtained from the similarity among vectors representing each phase.

3.2 Formulation of Our Method

In this section, we present details and formulation about each function shown in Fig. 1.

3.2.1 Conversion to Text Data

In order to analyze speech in meeting, it is necessary to convert speech data to text data. In this paper, we use human transcription for conversion. For our future work, we consider applying auto voice recognition.

3.2.2 Division into Phases

This function divides an entire text into n phases as follows:

Step 1: Count the number of characters m in text.
Step 2: Calculate the number of baseline l by dividing m by n.
Step 3: Divide an entire text into sentences.
Step 4: Construct one phase.

The system counts the number of characters with a sentence as a unit, and when the total number of letters exceeds 1, the system constructs one phase.

In these steps, we can apply the other text segmentation methods such as TextTiling [7] for dividing into each phase. In our method, we focused on time series. We divide a text based on number of letters, because we assuming that the number of letters is proportional to elapsed time. In our future work, we will be able to divide a text based on them by adding a time stamp to each speech in transcription.

3.2.3 Vectorization

This function consists of two steps: "Extract words (Morphological analysis)" and "TF-iPF (term frequency-inversed phase frequency) weighting." By these 2 steps, the system generates a vector par each phase. The vector's elements are featured by words in an entire text. The vectors' elements are weighted by TF-iPF weighting algorithm. We proposed this algorithm with reference to the TF-IDF algorithm.

(a) Word Extraction (Morphological analysis)

We need to obtain a list of base form of occurrence words in a text. In this paper we analyze Japanese text. In Japanese text, words are not separated. The system extracts all words from a text data by morphological analysis. We adopt only noun as an element of the vectors. Each vector's elements are featured by these words. Dimension of the vectors are the number of the extracted words.

(b) TF-iPF (term frequency—inversed phase frequency) weighting

We propose TF-iPF weighting algorithm with reference to the TF-IDF algorithm [11]. In our algorithm the phase corresponds to the document in TF-IDF. The vectors' elements are weighted by this scheme. TF-iPF algorithm is shown as belows:

$$tfipf_{i,j} = tf_{i,j} \cdot ipf_i$$
$$tf_{i,j} = \frac{a_{i,j}}{\sum_k a_{k,j}}$$
$$ipf_i = \log \frac{n}{|\{p: t_i \in p\}|}$$

Let $a_{i,j}$ be the number of occurrences of word i in the phase j · $\sum_k a_{i,j}$ is the sum of occurrences of words in all phases. Let $|A|$ be the number of elements of set A · $|\{p: t_i \in p\}|$ is the number of phases contains the word t_i.

3.2.4 Extraction of Feature Words in Each Phase

Feature words are extracted from the vectors by adopting most weighted k elements in the vector.

3.2.5 Calculation of Similarity Among Phases

Similarities among the vectors of phases are calculated by cosine measure. The result becomes a correlation matrix of $n \times n$.

3.2.6 Clustering of Phases

This function clusters phases based on similarities among phases. We regard each vector as a node on the graph. Let ε be threshold about connection between nodes. Each pair of node obtains connection with an edge when the similarity of the pair is less than ε. Let a group of node connected by the edges be a cluster. Find ε that maximizes the number of clusters. The clustering result is utilized for finding ε which maximizes the number of clusters. Finally, the function forms some groups including some phases.

3.2.7 Extraction of Topic Words in Each Phase

This function obtains topic words from the group to which the phase belongs and feature words in each phase. The function picks up the words which commonly exist in the phases belonging to the same group. In these words, the function extracts the words as topic words by selecting words appearing in more phases. The chosen word is a word which appears in at least two or more phases. The number of topic word can be set arbitrarily. In this paper, we set the maximum number of topic word to 20.

3.2.8 Calculation of Importance in Each Phase

This function calculates the importance of each phase from the $n \times n$ correlation matrix expressing the relationship among the phases. The value of each element of the matrix is the cosine similarity of the two phases. One column of the matrix represents the similarity between one phase and the other phases. The function calculates the sum value of each column. The phases having the large sum values are much related to other phases in the conference and have a central topic. The function calculates the sum of the columns in all phases and normalizes them with an infinity norm. Let the normalized values be the importance of the phases. By normalization, the importance of the phase with the highest importance becomes 1, and the importance of the meeting is represented relatively from it.

3.3 Three Output Visualizations

Our system finally outputs three visualization, "Phase feature words," "Meeting-outline," and "Importance graph."

(1) Phase feature words
A user selects an arbitrary phase. The system presents the feature words of the selected phase. When the phase belongs to any group, the system presents both the topic word and the feature word of the group. When the phase does not belong to any group, the system presents only feature words. A user can use the Phase feature words to check contents of the phases in more detail.

(2) Meeting-outline
Meeting-outline is visualization for reviewing the flow of the conversation as the overview of the meeting. The system arranges each phase using time series as a coordinate axis. Each phase is classified by color according to the groups formed by the "Clustering of phases" function shown in 3.2.6. We can find the rough flow of topics of the meeting by the time series coloring.

(3) Importance graph
Importance graph is arranging the importance of the phases on time series. By focusing on phases of high score in the Importance graph, a user can figure out the phases including the central topic in the meeting. Conversely, by focusing on phases of low score in the "Importance-Graph", a user can also figure out latent issues of the meeting.

4 Preliminary Experiment

In this paper, we present a preliminary experiment. We implemented our proposed method and construct the meetings analysis system. We apply it to two actual meetings.

4.1 *Experiment System Setting*

An experiment system is implemented with the configuration shown in Fig. 1 by R and Python.

In this experiment, we use text data of actual meetings transcribed by human. For our future work, we will apply auto voice recognition. The type of meetings is a planning meeting. Participants in a meeting are talking in Japanese.

In this experiment, we show two analysis results of meetings by different members on different days. Table 1 shows the number of characters (in Japanese), sentences, members, and the total time of each meeting as an outline of these two meetings.

4.2 *Experiment Result 1: In the Case of Meeting-1*

Figure 2 shows the Meeting-outline and the list of topic words in the Meeting-1. The Meeting-outline shows which group each phase belongs to. In addition, the list of topic words shows topic words of each group. "****" in the table is a word which we cannot disclose such as a company name, a personal name, etc.

In this result, three groups are formed. The group 1 has the most number of phases. Phases 6–12 belong to group 1 continuously. Phase 1 also belong to group 1. Phase 3–4 and phase 20 belong to group 2. Phase 13 and 18 belong to group 3. Phase 2, 5, from 14 to 17, and 19 do not belong to any group. We can assume that the discussion was confused there.

In the group 1, there are "region", "area", or several words related to them. In the group 1, we can find that it is mainly talked about finding the relationship between area and attributes of people. That is, group 1 is a topic of people's area and attributes.

Table 1 Outline of two meeting for experiment

	Meeting-1	Meeting-2
Number of characters (Japanese)	15,553	21,807
Number of sentences	323	327
Number of members	6	6
Total time	2 h 18 m	1 h 42 m

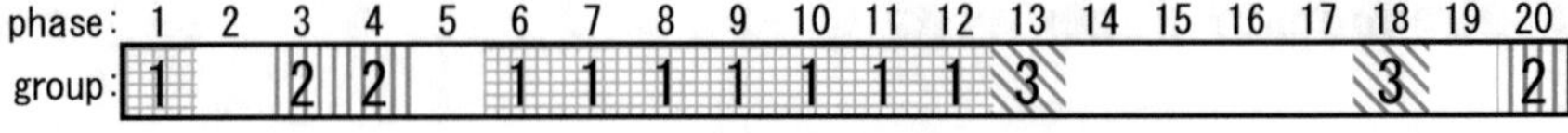

1	2	3
region	room	time
position	space	health
****	office	log
life	registration	behavior
area	ICT	comic
attribute	SNS	ability
res	meeting	company
property	provision	
hobby	liking	
concept of values	hobby	
health	device	
downtown	customer	
house	account	
buying and selling	booking	
opulence		
prediction		

Fig. 2 Meeting-outline and the list of topic words in meeting 1. Meeting-outline shows which group each phase belongs to. In addition, the list of topic words shows topic words of each group. In this result, three groups are formed. The group 1 has the most number of phases. Phases 6 to 12 belong to group 1 continuously. We can make a conjecture that the group contains the most important topic. On the other hand, phases 14 to 17 do not belong to any group. We can assume that the discussion was confused there

In the group 2, there are words such as "office", "hobby", "SNS", "account". We can find that it is mainly talked about data that can be obtained from human activities. That is, group 2 is a topic of human activity data.

In the group 3, there are words such as "comic", "health", "log". We can find that it is mainly talked about human activity log. Especially, it is talked about logs related to "comic" and "health" applications. That is, group 3 is a topic of comic and health application data.

In this result, we can overlook the flow of conversation in the Meeting-1. First, they start talking about a topic of people's area and attributes. Next, they talk about a topic of human activity data. In the middle of the meeting they talk about the topics of people's area and attributes intensively. After that, the content of the discussion is chaotic. Finally, they talk about a topic of human activity data again and it has ended the meeting.

Figure 3 shows Importance graph of Meeting-1. In the Fig. 3, we can find more important phases and less important phases. We can easily find the important phase and understand the main points of discussion in the meeting. In addition, by

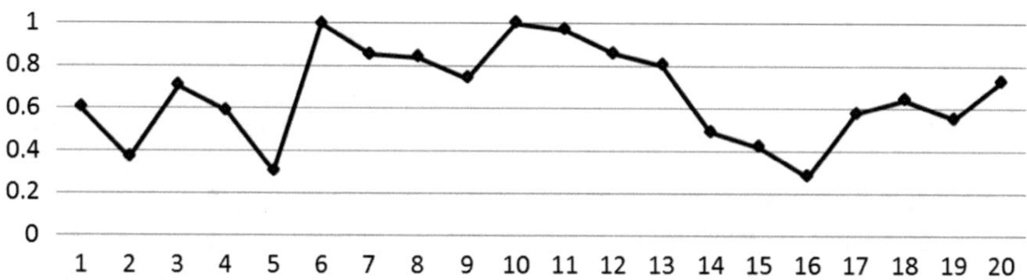

Fig. 3 Importance graph of meeting-1. We can find more important phases and less important phases. We can easily find the important phase and understand the main points of discussion in the meeting. In addition, by focusing on less important phases, we can find latent issues of the meeting and we may find reasons what prevent a process of the meeting. We can find that the most important phase is the phase 6 and the least important phase is the phase 16 in this graph

focusing on the less important phases, we can find latent issues and find reasons what prevent a process of the meeting.

We can find that the most important phase is the phase 6 and the least important phase is the phase 16 in this graph. It can be seen that from phase 3 to 13 was a period of prosperity.

Table 2 shows two examples of Phase feature words of Meeting-1. This table shows phase 6 and phase 16 of Phase-Feature words, phase 6 is the most important phase and phase 16 is the least important phase in the above result. Phase 16 does not belong to any group. Phase 16 has only feature words in this table. In the phase 6, there are also "prediction", "user", "client", "consciousness", etc.

Table 2 Phase feature words of phase 6 and 16 in the meeting-1

Phase 6		Phase 16
Topic words	Feature words	Feature words
Region	Client	Software
Position	Icon	Biotech
Attribute	User	Dictionary
Res	Student	Algorithm
Hobby	Comic	Station name
Buying and selling	Taste	Processing
Prediction	B to B	Fluctuation of description
	B to C	Analysis
	Word	All
	Consciousness	****
	Meaning	Bullshit
	Difference	
	General	
	Perfect	
	Classification	

4.3 Experiment Result 2: In the Case of Meeting-2

Figure 4. shows Meeting-outline and the list of topic words in Meeting-2. How to read the graph is the same as in Fig. 2. In this result, four groups are formed. The group 3 has the most number of phases. Group 2 has the second most number of phases. We can infer that group 3 and group2 are relatively important groups. On the other hand, phases 8–11 do not belong to any group. We can assume that the discussion was confused there.

In this result, we can overview the flow of conversation in the Meeting-2. Actually, Meeting-2 is a meeting on this dialogue structure analysis system. First, they talk about a basic function of this system. In the Fig. 4, the group 1 corresponds to it. Next, they talk about how to use this system for a meeting facilitation. In the Fig. 4, the group 2 corresponds to it. Furthermore, they talk about how to estimate this system. In the Fig. 4, the group 3 corresponds to it. Finally, they summarize the contents of the meeting. In the Fig. 4, the group 4 corresponds to it.

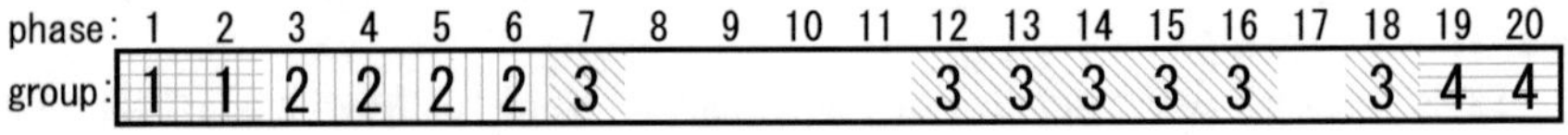

phase:	1	2	3	4	5	6	7	8	9	10	11	12	13	14	15	16	17	18	19	20
group:	1	1	2	2	2	2	3					3	3	3	3	3		3	4	4

1	2	3	4
conversation	production	conference	word
self	meeting	decision	phase
whole	field	minutes	events
****	speech	intention	estimate
cosine	table	time	simple
scale	mechanism	committee	time series
system	model	topic	log
time	meaning	discussion	common
search	image	boss	myself
click	indicator	research	****
common	room	common	
	voice	method	
	facilitator	normal	
	alert	vocabulary	
	monitor	content	
	support	pattern	
	question	meeting	
	****	plan	
	facilitation		

Fig. 4 Meeting-outline and the list of topic words in meeting-2. Four groups are formed. Group 3 has the most number of phases. Group 2 has the second most number of phases. We can infer that group 3 and group 2 are relatively important groups. On the other hand, phases 8 to 11 do not belong to any group. We can assume that the discussion was confused there

Based on this result, by grouping phases our system realizes to express the flow of conversation in the meeting.

Figure 5 shows Importance graph of Meeting-2. The most important phase is the phase 6. The least important phase is the phase 8 in this graph. When we would like to find a hot topic of Meeting-2, we only check the Phase feature words of phase 6. Table 3 shows a part of the Phase feature words of phase 6. By these Phase feature words shown in Table 3, we can find that how to use this system during brainstorming is the one of the most important issue in Meeting-2.

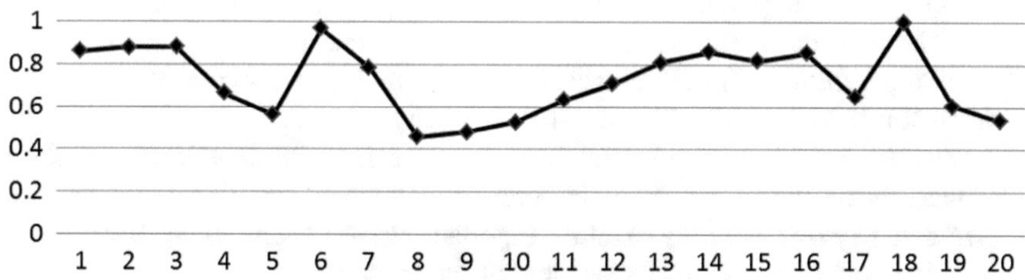

Fig. 5 Importance graph of the meeting-2. We can find that the most important phase is the phase 6 and the least important phase is the phase 8 in this graph

Table 3 Phase feature words of phase 6 and 16 in the meeting-2

Ph. 06	
Topic words	Feature words
Production	Brainstorming
Meeting	Floor
Field	"How to use"
Speech	Abundance
Table	Threshold
Mechanism	Vocabulary
Model	Intention
Meaning	Decision
Image	****
Indicator	****
Room	Output
Voice	Input
Facilitator	Dashboard
Alert	Template
Monitor	
Support	
Question	
****	
Facilitation	

5 Conclusion

We presented a dialogue structure analysis method to visualize the transition of topics in a meeting as the one of dialogue process representation. Our method extracts topics in a meeting on time series. In addition, we define an index to assess the importance of the whole meeting in each phase.

We proposed three visualizations, "Phase feature words," "Meeting-outline," and "Importance graph." The Phase feature words shows important keywords and topics. The Meeting-outline shows the flow of the conversation as the overview of the meeting. The Importance graph shows the importance of the phases on time series. Through these three visualizations, you can understand the excitement and flow of the meeting. These visualizations help to know process of arriving at decision for dialogue by facilitating and reviewing a meeting.

Moreover, we presented a preliminary experiment. In the experiment, we show two analysis results of meetings by different members on different days. The implemented system not only extracts important keywords and topics but also the flow of the conversation as the overview of the meeting.

In the near future, our method will be applied to a meeting facilitation. We must consider how to visualize the flow of the conversation in real time.

References

1. NTT DATA Institute of management consulting: Survey on 'conference innovation and work style' (in Japanese). https://www.keieiken.co.jp/aboutus/newsrelease/121005/;retr.2017/3/23
2. Cameron, D.: Working with spoken discourse. SAGE Publications Ltd (2001)
3. DiMicco, J.M., Hollenbach, K.J., Pandolfo, A., Bender, W.: The impact of increased awareness while face-to-face. Hum. Comput. Interact. **22**(1–2), 47–96 (2007)
4. Bergstrom, T., Karahalios, K.: Conversation clock: visualizing audio patterns in co-located groups. In: Proceedings of the 40th Annual Hawaii International Conference on System Sciences (HICSS '07), p. 78. IEEE Computer Society, Washington, DC, USA (2007)
5. Olguín, D.O., Waber, B.N., Kim, T., Mohan, A., Ara, K., Pentland, A.: Sensible organizations: technology and methodology for automatically measuring organizational behavior. IEEE Trans. Syst. Man. Cybern. Part B (Cybernetics) **39**(1), 43–55 (2009)
6. Blei, D., Ng, A., Jordan, M.: Latent dirichlet allocation. J. Mach. Learn. Res. 1107–1135 (2003)
7. Hearst, M.A.: TextTiling: segmenting text into multi-paragraph subtopic passages. Comput. Linguist. **23**(1), 33–64 (1997)
8. Choi, F.Y.: Advances in domain independent linear text segmentation. In: Proceedings of the 1st North American Chapter of the Association for Computational Linguistics Conference, pp. 26–33. Association for Computational Linguistics (2000)
9. Riedl, M., Biemann, C.: How text segmentation algorithms gain from topic models. In: Proceedings of the 2012 Conference of the North American Chapter of the Association for Computational Linguistics: Human Language Technologies (NAACL HLT '12), pp. 553–557. Association for Computational Linguistics, Stroudsburg, PA, USA (2012)

10. Riedl, M., Biemann, C.: TopicTiling: a text segmentation algorithm based on LDA. In: Proceedings of ACL 2012 Student Research Workshop (ACL '12), pp. 37–42. Association for Computational Linguistics, Stroudsburg, PA, USA (2012)
11. Baeza-Yates, R.A., Ribeiro-Neto, B.A.: Modern information retrieval: the concepts and technology behind search, 2nd edn. Addison-Wesley Professional (2011)

Joint Embedding of Hierarchical Structure and Context for Entity Disambiguation

Shuangshuang Cai and Mizuho Iwaihara

Abstract Entity linking refers to the task of constructing links between the mentions of context and the description pages from knowledge base. Due to the polysemy phenomenon, the key issue of entity linking is entity disambiguation. To simplify the goal of entity disambiguation, the main problem is choosing the correct entity from candidates. In this paper, we propose a novel embedding method specifically designed for entity disambiguation. Existing distributed representations are limited in utilizing structured knowledge from knowledge bases such as Wikipedia. Our method jointly maps the information from hierarchical structure of knowledge and context words. We extend the continuous bags-of-words model by adding hierarchical categories and hyperlink structure. So far, we have trained a joint model which adds category information. We demonstrate the utility of our proposed approach on an entity relatedness dataset. The results show that our jointly embedding model is superior to the model simply using context words. In addition, we do disambiguation experiment on a dataset, and the results show slight superiority of the novel embedding model.

Keywords Entity disambiguation · Categorical hierarchy · Hyperlinks · Semantic relatedness

1 Introduction

With the development of the Internet, more and more texts can be approached. It is important to provide assistance in finding desired information from a vast amount of text information. To this end, various text mining tasks have been researched,

S. Cai (✉) · M. Iwaihara (✉)
Graduate School of Information, Production and Systems, Waseda University, Kitakyushu 808-0135, Japan
e-mail: shuangshuang@ruri.waseda.jp

M. Iwaihara
e-mail: iwaihara@waseda.jp

R. Lee (ed.), *Software Engineering, Artificial Intelligence, Networking and Parallel/Distributed Computing*, Studies in Computational Intelligence 721,
DOI 10.1007/978-3-319-62048-0_5

including document summarization, text classification and text clustering. Meanwhile, entity-centric data has led researchers to a new direction of constructing an entity-based network, where Google Knowledge Graph is a well-known example. For web text data, the knowledge bases which contain rich knowledge about words and contexts can help to solve named entity disambiguation problem [10] which is to find a correct entity mentioned by its name in the context.

Tweets, blog postings and new articles contain mentions of entities, where these mentions are often indicating entities such as people, companies, organizations and places. To help readers easily understand the context and decrease reading difficulties from polysemy and synonymy, such mentions need to be linked to a corresponding descriptive entity page. However, mentions are sometimes ambiguous, such that a name may refer to different entities in different contexts. For example, a mention of Washington can be linked to dozens of possible entities in Wikipedia, such as George Washington, Washington, D.C., the University of Washington, etc. This ambiguity problem causes the difficulty of entity linking.

Let us see another example. Given a sentence "Washington is a state in the Pacific Northwest region of the United States located north of Oregon". How can a computer distinguish that "Washington" is a state in the United States, not the president and the word "state" denotes a U.S. state rather than a state of matter in chemistry? The purpose of entity linking is to detect mentions and find their referent entities from a knowledge base (e.g., Wikipedia). Figure 1 shows the mapping result of entity linking.

Wikipedia is a multilingual free internet encyclopedia. There are more than five million articles in English Wikipedia. For this reason, Wikipedia is often used as a knowledge base for entity disambiguation. In entity disambiguation, one Wikipedia article is regarded as an entity, its article title is regarded as the entity name, and the article content is regarded as an entity description or context.

Distributed representations of words were proposed early, but recently have been successfully applied to language models and several natural language processing

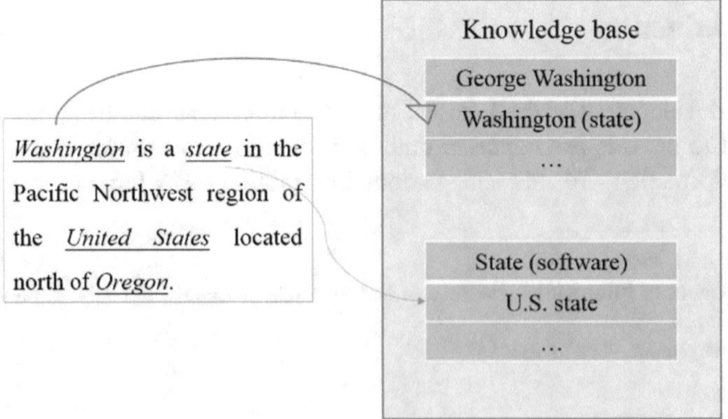

Fig. 1 Illustration of entity disambiguation task

tasks, including word embedding [9, 14]. Traditional word representations are based on a co-occurrence matrix, which suffers from high dimensionality. Recent word embedding has brought new ideas to this area. We discuss utilizing word embedding for measurement of semantic relatedness in entity linking, but its straightforward application is not considering linkage between articles.

In this paper, we propose a novel word embedding that jointly use the information from the context and the hierarchical structure of Wikipedia.

Rich categories and inter-article links are remarkably interesting aspects of Wikipedia [7, 10]. Categories in Wikipedia form a directed graph, having the root category called the main topic classification, which can be backtracked from each non-root category. Each category can have an arbitrary number of subcategories, and similarly one category may have more than one parent categories. One article is assigned to one or more categories during users' editing. Categories act as a semantic tag, and articles that have similar topics usually belong to the same category.

The links between Wikipedia articles also contain numerous information. One ambiguous entity will be linked to a referent article so that people can easily distinguish the ambiguous entities. If two Wikipedia articles have more internal links to common articles, then these articles share more common related contents, so we can assign a high relatedness score to such article pairs.

Our contributions of this paper can be summarized as follows:

- We extend the CBOW model [8] to reflect contexts arising from the structure of Wikipedia. The hierarchical structure of categories and links are utilized to capture related words, which can enrich prediction reference of the word embedding model. Our model integrates all information together to learn the distributed representation of words without supervision.
- After training the novel word embedding model on the whole Wikipedia, the model can be directly used on entity disambiguation task, as well as other tasks based on semantical analysis.

The rest of this paper is organized as follows: Sect. 2 gives an overview of various approaches to word embedding and related work on entity disambiguation. Section 3 describes our proposing methods. Section 4 presents two experiments for evaluating relatedness between entities and entity disambiguation, comparing the effects of different models, and making an analysis of the results. The paper concludes in Sect. 5.

2 Related Work

Wikipedia is a quite useful information resource of natural language processing, with various derived huge knowledge, such as DBpeida [1], Freebase [2], and BabelNet [11]. Document summarization, text classification and text clustering can take advantage of these resources. When it comes to entity disambiguation, the problem is addressed from different perspectives. Using link-based structure and using text in articles are two main Wikipedia-based approaches.

Milne and Witten [15] assessed semantic relatedness. They computed the number of incoming links that the two articles have common based on Normalized Google Distance (NGD) on hyperlinks between articles. Nevertheless, NGD can miss links that should be given, but missed or omitted during editing articles. Diego Ceccarell [4] provided a number of features to measure relatedness for entity linking. They obtained results that their machine-learned entity relatedness function performed better than other functions.

A distinct direction in entity disambiguation focuses on the effects of mention's context. The relevant entities of a mention can be derived from its context. TF-IDF measure and cosine similarity are often used on bag-of-words. The category graph of Wikipedia is widely used as well. Torsten and Iryna [18] analyzed the Wikipedia category graph for NLP application. In [7], the authors performed a graph-theoretic analysis of the Wikipedia category graph and showed that the graph is well suited to estimate semantic relatedness between words.

Word embedding methods are also becoming increasingly popular since it was published in 2013 [8]. The main advantage of word embedding is that the word representation of two similar words are very close in the vector space and the dimension of words can be decreased significantly in comparison with the traditional bag-of-words model.

Relational embedding is a family of methods to represent words as vectors and relations as operations applied to words such that certain properties are preserved [12]. For instance, the linear relational embedding [5, 13] applies a relation to an entity based on matrix-vector multiplication, while TransE [3] simplifies the operation to vector addition. Yamada [16] proposed a novel embedding method specifically designed for entity disambiguation. This method jointly maps words and entities into the same continuous vector space. To capture structured semantics, Hu proposed a principled framework of embedding entities that integrates hierarchical information from large-scale knowledge bases [17].

Our approach is based on our former research [19] and we propose a new way to jointly use the hierarchical categories and links to reflect the linking contexts of Wikipedia into word embedding.

3 Proposed Method

In this section, we extend the classical continuous bog-of-words model (CBOW) model for learning word embedding. Information from contexts, hierarchical categories and links from knowledge bases will be combined to map a word to a continuous d-dimensional vector space.

3.1 Continuous Bag-of-Words Model (CBOW)

The training criterion of the CBOW model is to predict the middle word by building a log-linear classifier with N future and N history words. The order of words does not influence the projection. Formally, given a sequence of T words $w_1, w_2, w_3, \ldots, w_T$, the model aims to maximize the following objective function:

$$L_w = \sum_{t=1}^{T} \sum_{-c \le j \le c, j \ne 0} \log P(w_t | w_{t+j})$$

where c is the size of the context window, w_t is the target word and w_{t+j} denotes the contextual words besides the target word. The conditional probability $P(w_t | w_{t+j})$ can be calculated by the following softmax function:

$$P(w_t | w_{t+j}) = \frac{\exp(v_{w_t}^{'T} v_{w_{t+j}})}{\sum_{w \in W} \exp(v_{w_t}^{'T} v_w)},$$

where W contains all the words in the vocabulary, $v_w^{'}$ and v_w denote the vectors of w in matrices $v_w^{'}$ and v_w, respectively. We train the CBOW model to optimize the above objective function. The output is the vector representations of words.

3.2 Extension of CBOW Model

In a structural knowledge base, its hierarchical categories and links also contain semantic information of a word. They can be incorporated in the word prediction to construct word vectors. However, most of the relations are based on entities rather than words. For this reason, we simply extend words to entities. We regard the Wikipedia page titles as entities.

3.2.1 Hierarchical Categories

Unlike words that can be directly extracted from context, entities are hidden under their surface occurrences. Articles in Wikipedia have a hierarchical structure. As we know, articles with common parent categories usually have high semantic relatedness.

An example is shown in Fig. 2, where article Apple Inc. belongs to several categories, such as Apple Inc. and Computer hardware companies. Meanwhile, category Apple Inc. has more than ten subcategories and more articles under it. Likewise, category Computer hardware companies has its subcategories and articles. These super-categories and subcategories constitute a graph representing the category hierarchy. Since articles under the same parent category share the topics of

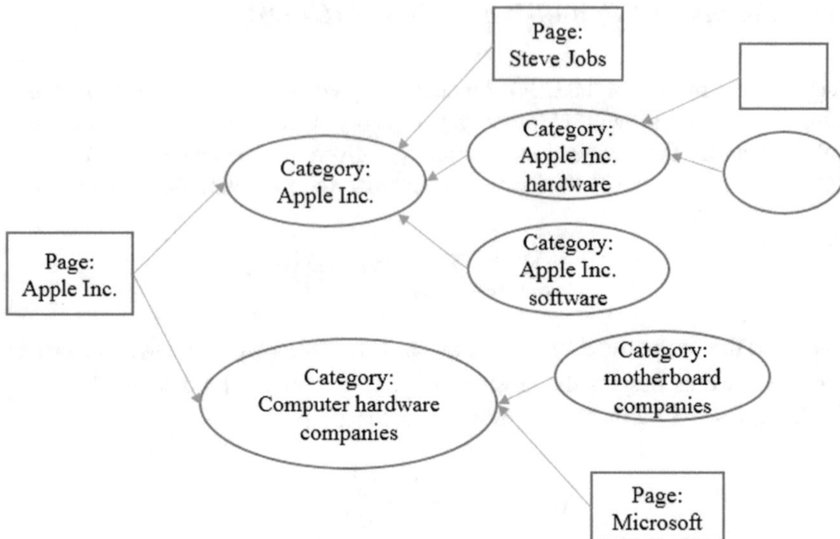

Fig. 2 Wikipedia category example

the category, they are expected to have high semantic relatedness. Therefore, we measure the relatedness between two articles using a distance based on the hierarchical category structure (Fig. 3).

In computing the distance on the graph, we consider entities as leaf nodes and categories as internal nodes. The measurement is based on the hypothesis that nearby entities tend to share common topics. Particularly, we define a distance metric $M_c \in R^{n \times n}$ for each category node. After that, we introduce an aggregated distance metric to measure the distance between two entities.

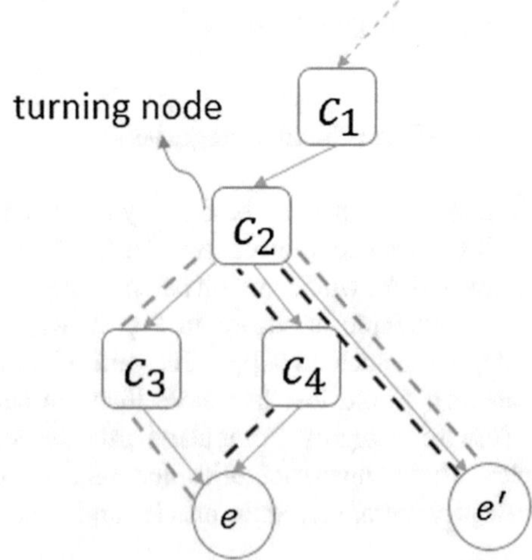

Fig. 3 Paths in hierarchical categories structure

Let $P_{e,e'}$ be the path between two entities e and e'. It is easy to define an aggregated metric $M_{e,e'} \in R^{n \times n}$ by summing the distances of all category nodes on the path. If there exist more than one path, we could enumerate all the existing paths, but the number of such paths is exhaustive and unrealistic.

To reflect all the related categories and avoiding intractable complexity, we extend $P_{e,e'}$ as the set of all category nodes contained in any of the paths between e and e'. Then we can define the aggregated relatedness:

$$M_{e,e'} = \gamma_{e,e'} \sum_{c \in P_{e,e'}} \pi_{ee',c} M_c$$

where $\{\pi_{ee',c}\}$ are the relative weights of the categories and $\sum_{c \in P_{e,e'}} \pi_{ee',c} = 1$. We can calculate $\pi_{ee',c}$ using the average steps descending from the category node c to entity e in the graph. The weight can be expressed as $\pi_{ee',c} \propto \left(\frac{1}{s_e} + \frac{1}{s_{e'}}\right)$. It can be simply understood as an entity is more relevant to its immediate categories than to its further ancestors. The factor $\gamma_{e,e'}$ measures the distance between the entities in the hierarchy. The distance can be measured in various ways. In this paper, we adopt it as $\min\{s_e, s_{e'}\}$.

After the aggregated relatedness is introduced, the distance between two entities can be measured as:

$$d(e, e') = (v_e - \bar{v}_{e'})^T M_{e,e'} (v_e - \bar{v}_{e'})$$

3.2.2 Wikipedia Link Structure

Wikipedia internal links connect one Wikipedia article to another, in contrast to external links which connect an article to an external website. We assume that entities with similar links are related (Fig. 4).

Anchors in Wikipedia article texts are terms or phrases to which links are attached. We use anchors to identify candidate articles for terms. Wikipedia's documentation dictates that any term or phrase that relates to a significant topic should be linked to the article that discusses it. Consequently, it provides a vast number of anchor texts which capture both polysemy and synonymy. For example, *plane* links to different articles depending on the context in which it is found, and sometimes *plane*, *airplane* and *aeroplane* are all used to link to the same article.

To simplify the problem, we only consider entities appearing in the same article. Usually, most important words or entities which have significant meaning are marked as anchors in Wikipedia. If two entities represented by two articles have common link destinations, we can regard that having such common link destinations can raise semantic relatedness of the two entities. The Wikipedia link structure possesses large amount of link information and it is greatly helpful for disambiguation of words and entities.

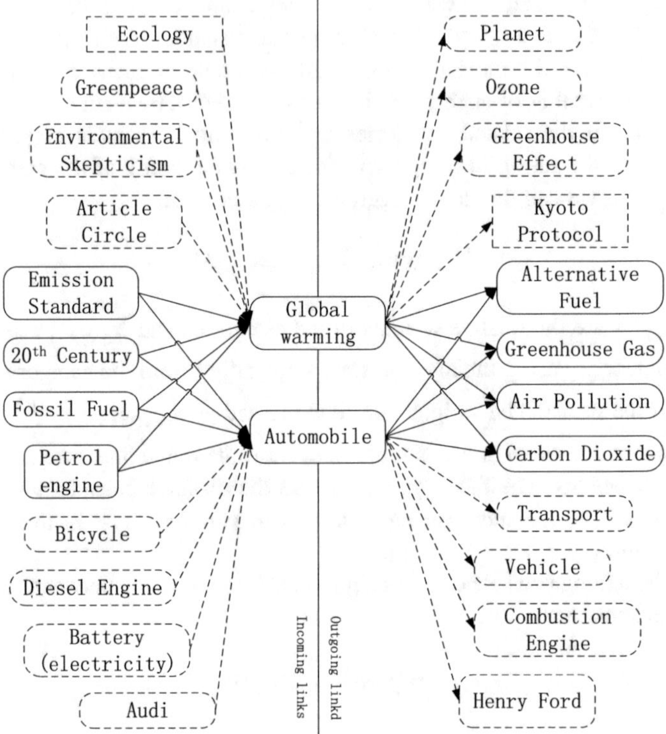

Fig. 4 A semantic relatedness measure between Automobile and Global warming from Wikipedia links

Based on the above assumption, the relatedness between two entities [10] can be computed using the following function:

$$R(a,b) = 1 - \frac{\log(\max(|A|,|B|)) - \log(|A \cap B|)}{\log(|W|) - \log(\min(|A|,|B|))},$$

where a and b are two entities, A and B are the sets of all articles that link to a and b, respectively, and W is the entire Wikipedia articles.

4 Experimental Evaluation

In this section, we first describe the details of training our joint embedding. Before disambiguation experiments, we first evaluate the quality of word embedding on entity relatedness.

4.1 Training

To train the joint embedding model, we used the October 2016 version of the Wikipedia dump. We first preprocessed the corpus by removing articles whose words were fewer than 50, and finally we obtained 4.1 million articles. We also removed words that appeared less than 5 times, and we obtained 2.1 million distinct words. We parsed the Wikipedia articles and extracted categories and anchors from each article. We separately computed the relatedness between two entities using hierarchical categories and link structure.

We currently incorporated the category measure into the word embedding model. The number of dimensions of word embedding was set to 100. The learning rate was set to 0.001, and the size of the window was 10. In consideration of window size, for common words and entities, we reserve five most related words or entities. The model was trained by iterating over articles in the Wikipedia dump five times. The training lasted approximately one day.

4.2 Entity Relatedness Experiment

4.2.1 Experimental Setting and Dataset

We implemented our proposed model, and evaluated our joint embedding model over an entity relatedness dataset, which trained by the English Wikipedia dump from October 2016.

To compare the quality of relatedness scores between entities obtained by our approach, we performed an experiment on entity relatedness. We use the KORE-relatedness-entity dataset from [6], in which the authors selected a set of 21 queries that correspond to 21 entities from knowledge base YAGO2. These entities are selected from four different domains: IT companies, Hollywood celebrities, video games, and television series (Table 1). For each of the 21 seed entities, they selected 20 candidates from the set of entities linked to the seed's Wikipedia article. The KORE dataset is composed of 420 entity-candidate pairs in total. The authors used a crowdsourcing platform to obtain the gold standard ranking of the 20 candidate entities for each seed entity. The rank indicates the semantic relatedness between seed entity and candidates. For example, let us consider seed entity Apple Inc. in the IT Company domain, in which Steve Jobs was ranked 1 and it is obvious that he is the most related, while Ford Motor Company was ranked 20, reflecting its least relatedness.

We compare our ranking results with the gold standard of human-ranked results by Spearman's rank correlation coefficient (ρ). The coefficient ranges from −1 to 1, assessing how well the relationship between two variables can be, where a perfect positive correlation is represented by the value 1, while a value 0 indicates no correlation and −1 indicates a complete negative correlation. From this sense, the

Table 1 Example of seed entities and gold-standard ranks of candidate entities from [6]

Seed entity	Ranked candidate entities
Apple Inc. (IT Company)	Steve Jobs (1), Steve Wozniak (2) Jonathan Ive (3), Mac Pro (4) … Sears (19), Ford Motor Company (20)
Angelina Jolie (Hollywood Celebrity)	Jon Voight (1), Brad Pitt (2) … Chip Taylor (8), Academy Awards (9) … 2005 Kashmir earthquake (20)
Grand Theft Auto IV (Video Game)	Niko Bellic (1) Grand Theft Auto (series) (2) … Brooklyn Bridge (14), Metacritic (15) … Mothers Against Drunk Driving (20)

correlation ρ of our system's results and gold-standard ranks should be a positive value, and the larger is the better.

4.2.2 Experimental Results

The experimental results for the KORE relatedness entity dataset are shown in Table 2, where the Spearman correlation results between the gold-standard rank and the ranking results generated by our methods are shown.

The KORE method is from the original paper [6], based on keyphrase overlap relatedness. We adapt this method as our baseline.

The WoE method is the original word embeddings. The WoCE method is computed by the embedding model that we trained with hierarchical categories.

We can easily find that the result of our novel word embedding method is better than the original one, except one domain on Television series. However, there is not a large gap between the results. This may be caused by the following two reasons. The first one is that the ranking of candidate entities is done by human judgement. Whether Mac Pro or Steve Jobs is more related to Apple Inc. is hard to say, depending on situations. Another reason is that we compute the cosine similarity of two vectors, one is the seed entity's context vector and the other one is the

Table 2 Spearman correlation of relatedness measures with human ranking

Domain	KORE (baseline)	WoE	WoCE
IT company	0.208	0.472	**0.485**
Hollywood celebrity	0.522	0.554	**0.586**
Video game	0.499	0.477	**0.502**
Television series	**0.426**	0.119	0.306
Average	0.414	0.405	**0.469**

candidate's context vector. These context vectors are the average of their word vectors, where different contexts can be mixed.

Overall, the experiment result shows that our word embedding method is an effective tool for evaluating the relatedness between two entities. Integrating hierarchical categories improved the performance of word embedding method by 0.103–0.187 points in the Spearman's rank correlation coefficient.

4.3 Disambiguation Experiment

4.3.1 Experimental Setting and Dataset

The similarity in word2vec is computed by the cosine similarity of two word vectors. The similarity of an input sentence with a candidate is computed from the mean word vectors of the input and the candidate entity, as follows:

$$score(input, candidate) = sim\left(\frac{\sum_{input}^{m} v_m}{|input|}, \frac{\sum_{candidate}^{w} v_w}{|candidate|}\right)$$

Here, *input* is an input query. Variable m is every word in the input. |input| is the number of words in *input*. If the article is long, we set a context window of 100 words around the mention. *Candidate* is a preprocessed Wikipedia article for one candidate from the mention's candidate set. Variable w is every word in this candidate set. |*candidate*| is the number of words in this candidate set.

We performed our disambiguation experiment on a subset of the AQUAINT corpus [20]. The AQUAINT corpus consists of newswire text data in English, drawn from three sources: the Xinhua News Service, the New York Times Service and the Associated Press Worldstream News Service. We remove repeated articles and mentions that are common words, and verified links to Wikipedia titles. At last our dataset contains 45 news articles and 370 mentions, among which 266 mentions have at least two candidates. There are about average 220 words and average eight mentions in every news article. For mentions which have too many candidates, we only keep the top 10 candidates ranked by their popularities in Wikipedia and if the true entity is not in the candidate set, we add the true entity to its candidate set.

4.3.2 Experimental Results

We measured the accuracy as the fraction of mentions that were correctly linked to Wikipedia articles. Our disambiguation experiment result is shown in Table 3. The second line is the original word embedding method (WoE). The third line is computed by the proposed model integrating the hierarchical categories (WoCE).

Table 3 Disambiguation results on Acquaint corpus

Method	Correctly linked mention (total 370)	Accuracy (%)
WoE	257	69.5
WoCE	280	75.7

We can see a notable improvement of 6.2 percent in accuracy after adding hierarchical categories.

5 Conclusion and Future Work

In this paper, we proposed a novel embedding method specifically designed for entity disambiguation. We extend the continuous bags-of-words model by adding hierarchical categories and link structure. We showed the utility of our proposed approach on the entity relatedness dataset. Incorporating hierarchical categories into word embeddings shows a significant improvement over the model which simply uses context words.

We plan to evaluate the effect of adding the link structure for enriching the protection window. We further plan to add existing standard entity disambiguation features and evaluate our model on a larger dataset.

References

1. Auer, S., Bizer, C., Kobilarov, G., Lehmann, J., Cyganiak, R., Ives, Z.: DBPEDIA: a nucleus for a web of open data. In: The Semantic Web, pp. 722–735. Springer, Berlin, Heidelberg (2007)
2. Bollacker, K., Evans, C., Paritosh, P., Sturge, T., Taylor, J.: Freebase: a collaboratively created graph database for structuring human knowledge. In: Proceedings of ACM SIGMOD International Conference on Management of Data, pp. 1247–1250, June 2008 (2008)
3. Bordes, A., Usunier, N., Garcia-Duran, A., Weston, J., Yakhnenko, O.: Translating embeddings for modeling multi-relational data. In: Advances in Neural Information Processing System, pp. 2787–2795 (2013)
4. Ceccarelli, D., Lucchese, C., Orlando, S., Perego, R., Trani, S.: Learning relatedness measures for entity linking. In: Proceedings of 22nd ACM International Conference on Information and Knowledge Management, pp. 139–148 (2013)
5. Hiton, E.: Learning hierarchical structures with linear relational embedding. In: Advances in neural Information Processing Systems 14: Proceedings of the 2001 Conference, vol. 2, p. 857. MIT Press (2002)
6. Hoffart, J., Seufert, S., Nguyen, D.B., Theobald. M., Weikum, G.: KORE: keyphrase overlap relatedness for entity disambiguation. In: Proceedings of the 21st ACM International Conference on Information and Knowledge Management. ACM, pp. 545–554 (2012)
7. Huang, E.H., Socher, R., Manning, C.D., Ng, A.Y.: Improving word representations via global context and multiple word prototypes. In: Proceedings of 50th Annual Meeting of the Association for Computational Linguistics: Long Papers-Volume 1, pp 873–882, July 2012 (2012)

8. Mikolov, T., Chen, K., Corrado, G., Dean, J.: Efficient estimation of word representations in vector space (2013). arXiv:1301.3781
9. Mikolov, T., Yih, W.T., Zweig, G.: Linguistic regularities in continuous space word representations. In: HLT-NAACL, pp. 746–751, June 2013 (2013)
10. Milne, D., Witten, I.H.: Learning to link with Wikipedia. In: Proceedings of 17th ACM Conference on Information and Knowledge Management, pp. 509–518, Oct 2008 (2008)
11. Navigli, R., Ponzetto, S.P.: BabelNet: The automatic construction, evaluation and application of a wide-coverage multilingual semantic network. Artif. Intell. **193**, 217–250 (2012)
12. Nickel, M., Tresp, V., Kriegel, H.: Factorizing YAGO: scalable machine learning for linked data. In: Proceedings of 21st International Conference on World Wide Web, pp. 271–280 (2012)
13. Paccanaro, A., Hinton, E.: Learning distributed representations of concepts using linear relational embedding. IEEE Trans. Knowl. Data Eng. **13**(2), 232–244 (2001)
14. Taieb, M.A.H., Aouicha, M.B., Hamadou, A.B.: Computing semantic relatedness using Wikipedia features. Knowl.-Based Syst. **50**, 260–278 (2013)
15. Witten, I., Milne, D.: An effective, low-cost measure of semantic relatedness obtained from Wikipedia links. In: Proceeding of AAAI Workshop on Wikipedia and Artificial Intelligence: An Evolving Synergy, July 2008. AAAI Press, Chicago, pp. 25–30 (2008)
16. Yamada, I., Shindo, H., Takeda, H.: Joint learning of the embedding of words and entities for named entity disambiguation (2016). arXiv:1601.01343
17. Yuezhang, L., Ronghuo, Z., Tian, T., Zhiting, H., Rahul, I., Katia, S.: Joint embedding of hierarchical categories and entities for concept categorization and dataless classification (2016). arXiv:1607.07956
18. Zesch, T., Gurevych, I.: Analysis of the Wikipedia category graph for NLP applications. In: Proceedings of TextGraphs-2 Workshop (NAACL-HLT 2007), pp. 1–8, Apr 2007 (2007)
19. Zhang, Y., Iwaihara, M.: Evaluating semantic relatedness through categorical and contextual information for entity disambiguation. In: Proceedings of Computer and Information Science (ICIS), 2016 IEEE/ACIS 15th International Conference, pp. 1–6, June 2016 (2016)
20. The ACQUAINT corpus, Linguistic Data Consortium. https://catalog.ldc.upenn.edu/docs/LDC2002T31/

Automatic Optimization of OpenCL-Based Stencil Codes for FPGAs

Tsukasa Endo, Hasitha Muthumala Waidyasooriya and Masanori Hariyama

Abstract Recently, C-based OpenCL design environment is proposed to design FPGA (field programmable gate array) accelerators. Although many c-programs can be executed on FPGAs, the best c-code for a CPU may not be the most appropriate one for an FPGA. Users must have some knowledge about computer architecture in order to write a good OpenCL code. In addition, OpenCL-based design process requires several hours of compilation time, because re-writing and compiling many different OpenCL codes may require a very large design time. To solve this problem, we propose an automatic optimization method. We accurately predict the kernel performance using the log files generated at the initial stage of the compilation. Then we find the optimized FPGA architecture by searching all possible design parameters. We implement the proposed method to find the optimized architecture for stencil computation. According to the results, the design time has been reduced to 6–11% of the conventional approach.

Keywords OpenCL for FPGA · Performance tuning · Stencil computation · Code optimization

1 Introduction

FPGAs (field programmable gate arrays) are reconfigurable devices, where the user can change the architecture by a program. Usually, FPGAs are programmed using hardware description languages (HDL) such as Verilog or VHDL. However, this

T. Endo · H.M. Waidyasooriya (✉) · M. Hariyama (✉)
Tohoku University, 6-3-09, Aramaki-Aza-Aoba, Aoba, Sendai, Miyagi 980-8579, Japan
e-mail: hasitha@tohoku.ac.jp

M. Hariyama
e-mail: hariyama@tohoku.ac.jp

T. Endo
e-mail: tsukasa28@dc.tohoku.ac.jp

R. Lee (ed.), *Software Engineering, Artificial Intelligence, Networking and Parallel/Distributed Computing*, Studies in Computational Intelligence 721,
DOI 10.1007/978-3-319-62048-0_6

process is very time consuming and the programmer must have an extensive knowledge about the hardware design. Moreover, clock cycle-based simulations, timing and critical path analysis is required to efficiently reconfigure and FPGA. In addition, managing I/O resources is also a very difficult process. Users have to write hardware programs to use I/O controllers. Moreover, users also have to write device drivers and communication software in order to communicate and transfer data between an FPGA and a host CPU.

Recently, OpenCL for FPGA [1] is introduced to solve these problems. OpenCL is a framework to write programs to execute across heterogeneous parallel platforms [3]. It views a system as a number of computing devices (OpenCL devices) connected to a host. The host is usually a CPU while the devices can be any of OpenCL capable CPUs, GPUs, FPGAs, etc. A device contains one or more compute units and such compute units contain one or more processing elements. Kernels are the functions that are executed on an OpenCL device. The same OpenCL kernel code is executed on different OpenCL-capable devices such as CPUs, GPUs and FPGAs. The "offline compiler (AOC)" convert an OpenCL code into HDL and then into an FPGA bit-stream. It supports different FPGA boards by using a BSP (board support package) that contains the I/O and on-board memory information. OpenCL for FPGA contains software for data transfers and also the device drivers.

The same kernel program can be executed on different FPGA boards by recompiling it using the appropriate BSPs. However, different FPGAs and boards have different resources. Therefore, the optimum kernel code for one FPGA board may not be the best one for another board. Although many c-programs can be executed on FPGAs, the best c-code for a CPU may not be the most appropriate one for an FPGA. Users must have some knowledge about computer architecture in order to write a good kernel program. Otherwise, the designed kernel may not exploit the full potential of the FPGA. In addition, OpenCL-based design process requires several hours of compilation time, because re-writing and compiling many different OpenCL codes may require a very large design time.

In this paper, we proposed an automatic optimization method for OpenCL kernels. Our method is based on near-accurate prediction of the performance of an OpenCL kernel by analyzing the log files and resource utilization reports. Compilation of an OpenCL kernel can be divided into two stages: OpenCL to HDL code generation and HDL to FPGA bit-stream generation. We use the log files of the first stage compilation that takes significantly smaller compilation time compared to that of the second stage. An optimization problem is considered under resource constraints to find the optimal design for any OpenCL-capable FPGA board. We applied the proposed method for stencil computation kernels and achieved the optimized architecture in 6–11% of the design time compared to conventional methods.

2 OpenCL-Based FPGA Accelerator Design

2.1 Accelerator Design-Flow

The conventional OpenCL-based FPGA accelerator design-flow is shown in Fig. 1. It can be divided into three main phases. The emulation phase starts with a kernel program written in OpenCL. We can emulate the behavior of the kernel on a CPU to find that the desired outputs are achieved. Since the emulation is executed on a CPU, the processing time could be much larger than that of a CPU-oriented program. Therefore, designers usually use a scaled-down version of a program that use a very small data sample to get the output in a short time. If the program cannot be scaled-down, we have to skip the emulation phase, and directly go to the next phase. The next phase is the estimation phase. We compile the OpenCL kernel to get an HDL code, and we call it the "first stage compilation". After the first stage compilation, offline compiler provides an estimated resource usage report. Usually, there is a small difference between the estimated resource usage and the actual resource usage. If the resource usage is acceptable, we can proceed to the last phase. The last phase is the performance tuning phase. We compile the HDL code to generate a bit-stream

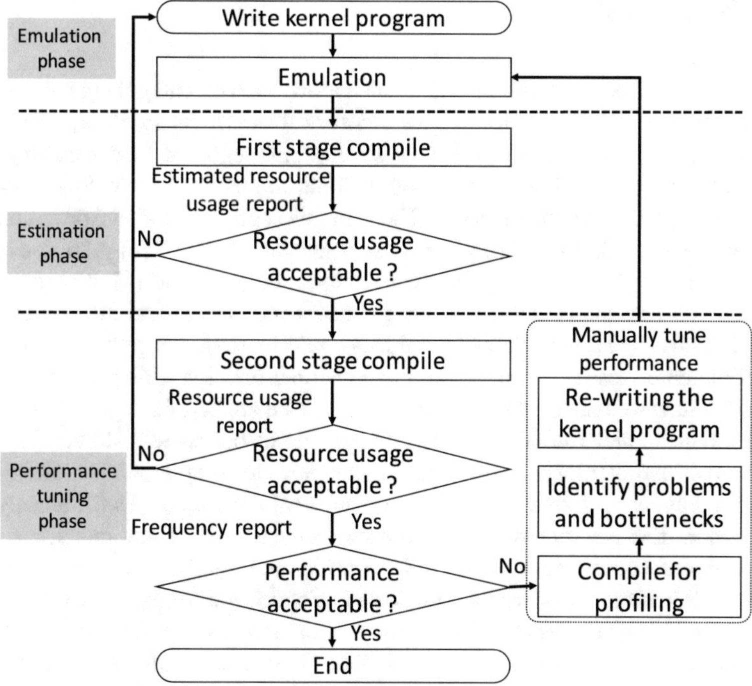

Fig. 1 Conventional OpenCL-based FPGA accelerator design flow

that is executable on an FPGA. We call this the "second stage compilation". The second stage compilation takes many hours of compilation time, since it involves time consuming processes such as placement, routing, etc. After the compilation, we get the actual area usage and the frequency information. Using those, we can determine whether the performance are acceptable or not. If the performance are not acceptable, we can identify the bottlenecks by recompiling the kernel for profiling. Then we can re-write the kernel to avoid the bottlenecks, and the whole process repeats again from the emulation phase.

As explained above, there is no concrete way for the user to know that the performance is optimal or even close to that. The profiling provides limited information about the bottlenecks of the kernel code. However, a different code could produce completely different performance. Therefore, the performance usually depends on the skill and the experience of the designer. Moreover, changing one part of the code could worsen the performance of the other parts. Therefore, it could be difficult to find how and which part should be corrected in order to increase the performance. In addition, the second stage compilation takes many hours of processing time. If we do this compilation often by re-writing the code, the design time could be hugely increased.

2.2 Previous Works

Since OpenCL-based FPGA design has been introduced recently, there are very few works propose better design techniques. Some of those works are [5, 7, 12]. Works in [5] considers several techniques such as using shift-registers, loop-unrolling, vectorization, using constant memory, using multiple compute units, etc. to increase the processing speed of the stencil codes. The work in [7] presents a visual programming framework called GALF-OCL, which is an extension of GLAF [6] proposed by the same authors. It also uses very similar techniques compared to [5] to optimize the OpenCL codes of molecular modeling, gene sequence search and filtering applications. In both [5, 7], considered techniques could increase the performance at the cost of increased resource and memory bandwidth utilization. One problem of these method is how to decide which technique should be applied for a particular FPGA board and an application. Since there are many FPGA boards with different amount of resources and memory bandwidths, we may have to compile many different kernels by applying various combinations of different techniques. Unfortunately, the compilation time is very large so that the kernel design time could increase significantly. Moreover, neither of these works consider an optimization problem. Therefore, it is difficult to know whether the kernel is optimized for the FPGA board.

Work in [12] proposes an optimization methodology for stencil codes. It calculates the processing time of an accelerator by a formula that contains several constant. The values of the constants should be measured by fully compiling several stencil computation codes. This could take a large design time. Moreover, determination of the constants should be done for different FPGA boards, different compilers and different stencil computation applications. As a result, this method could be inconvenient.

3 Proposed Automatic Optimization Methodology

There are two types of OpenCL kernels, single work-item kernels and NDRange kernels. The OpenCL codes of the NDRange kernels are quite similar to GPU kernels, and multiple work-items (corresponds to the threads in GPUs) are executed in a pipelined manner in an FPGA. The single work-item kernels follow a natural coding style similar to a typical c-program, where the loop-iterations are processed in a pipeline manner. In this paper, we consider only the single work-item kernels. However, similar approach could be used for NDRange kernels in future works.

3.1 Performance Prediction of the OpenCL Codes

Generally, OpenCL code of a single work-item kernel consists of multiple loops. Listing 1 shows an example of an OpenCL kernel code. It has a hierarchical loop structure, where the outer loop contains multiple inner loops. After doing the first stage compilation, we get a log file that contain information such as the loop number, "initiation interval (*II*)", pipeline inferring details, resource usage estimation, etc. Note that, *II* stands for the number of clock cycles between outputs. For example, if $II = 1$, an output is produced in every clock cycle.

```
__kernel void sample ( __global
const float * restrict A,
                          __global const float * restrict B
    ,
                          __global float * restrict C )
{
  float a, b;

  for(int i = 0; i < N; i++) {
    for(int j = 0; j < N; j++) {
      a += A[j];
    }

    for(int k = 0; k < M; k++) {
      b *= B[k];
    }

    C[i] = a + b;
  }
}
```

Listing 1 Loop structure of an OpenCL kernel

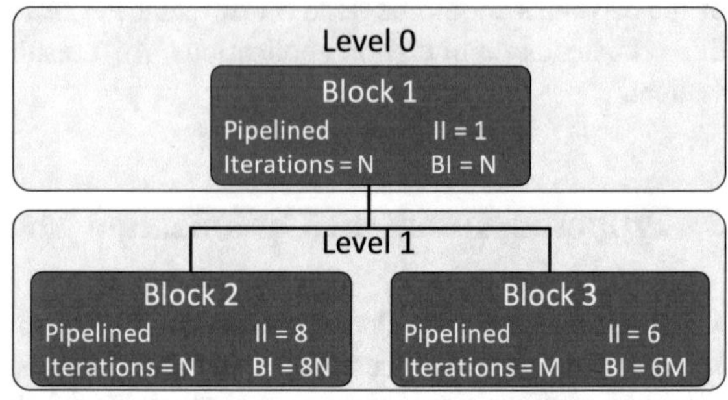

Fig. 2 The block structure corresponds to the kernel in Listing 1

We analyze the log file and the kernel code file to generate a block structure, where each block corresponds to a loop in the kernel code. Figure 2 shows such a block structure generated for the code in Listing 1. Each level in the block structure is associated with the loop hierarchy of the kernel code. A block is associated with a loop. The offline compiler tries to implement pipelines whenever possible. However, undetermined loop boundaries, global memory dependencies, mutual data dependencies among multiple loops, etc. could prevent pipeline implementation. The pipeline implementation details are also included in the block structure. The number of iterations is the loop-iterations given in the kernel code. The number of clock cycles required by the loop of a block is given by BI. It is defined by $BI = II \times$ iterations. Block information is used to predict the kernel performance.

If the parent block m of level t is denoted by p^t_m and its child blocks at level $t+1$ are denoted by $c^{t+1}_1 ... c^{t+1}_n$, the number of clock cycles required to execute the parent block ($p^t_m[cycle]$) is given by Eq. (1). Note that, $c^{t+1}_n[BI]$ is the number of clock cycles required by the loop of the child block c^{t+1}_n. When the parent block is pipelined, child blocks are executed in parallel so that the number of clock cycles depend on the largest BI of the child blocks. When the parent block is not pipelined, child blocks are executed in one by one so that the number of clock cycles depend on the sum of BI of all child nodes. This process continue for all parents from the bottom level to the top level (level 0). The total number of clock cycles required equals to the sum of the number of clock cycles in the parent nodes of level 0.

$$p^t_m[cycle] = \begin{cases} max\left\{c^{t+1}_1[BI], ..., c^{t+1}_n[BI]\right\} \times p^t_m[BI] & \text{if } p^t_m \text{ is pipelined} \\ \\ \sum_{k=1}^{n} c^{t+1}_k[BI] \times p^t_m[BI] & \text{if } p^t_m \text{ is not pipelined} \end{cases} \tag{1}$$

For example, the number of clock cycles of the kernel code in Listing 1 is given by follows, assuming $N = 100$ and $M = 200$.

$$max\,\{8 \times 100, 6 \times 200\} \times 100 = 120{,}000$$

Using this method, we can predict the number of clock cycles of any OpenCL kernel code. If we assume the clock frequency is constant, we can compare the number of

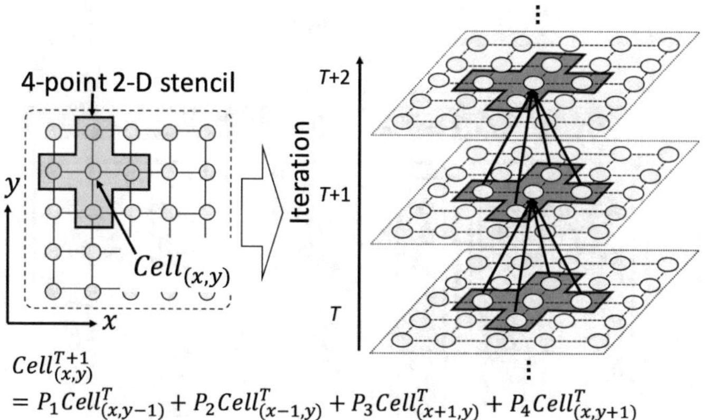

Fig. 3 Stencil computation using a 2-D 4-point stencil

clock cycles required by different codes and select the one with the minimum number of cycles as the best one.

3.2 Automatic Optimization for Stencil Computation Kernels

In this section, we use the proposed optimization methodology to implement stencil computation [9] kernels. Stencil computation is an iterative computation method used in many fields such as fluid dynamics [8], electromagnetic simulations [13], etc. It is a well studied problem and its FPGA oriented architectures have been proposed in many previous works such as [2, 10, 12]. However, there are many different stencil computation applications and many different FPGA boards, so that finding the optimal architecture for a given application and an FPGA board is a difficult and time consuming problem.

We explain the stencil computation architecture briefly. Figure 3 shows the computation of a 4-point 2-D stencil. The computations of the cells in a new iteration are done using the computation results of the cells in the previous iteration. Figure 4 shows the stencil computation architecture proposed in previous work [12]. It has multiple pipelined computation modules (PCMs) where each PCM processes one iteration. One stencil computation is done in one PE and multiple PEs are used to compute multiple stencils in parallel. Shift registers are used to carry forward the result of one iteration to the next PCM that computes the next iteration.

```
#define N ((HEIGHT*WIDTH +nPCM*(WIDTH+2))/nPE)

__kernel void stencil( global float * restrict
frame_in,
                       global float * restrict frame_out )
{
```

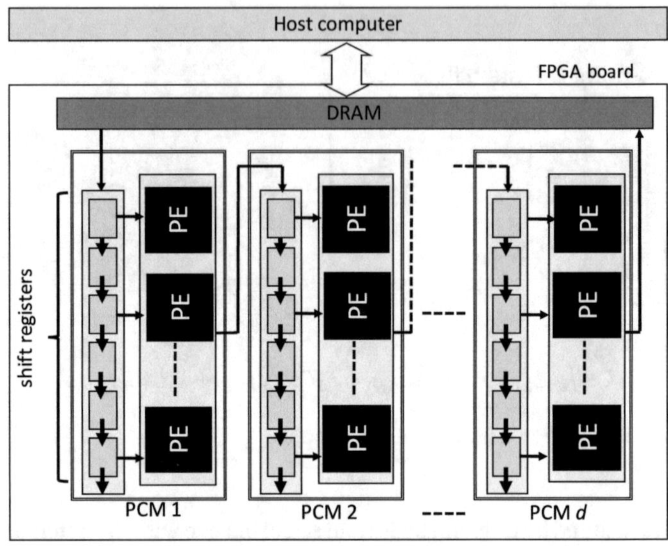

Fig. 4 Stencil computation architecture proposed in [12]

```
float rows[nPCM][...];

//manually unroll loop by nPE
for (int count=0, count<N, count++)
{
  #pragma unroll
  for (int i+..; i>0; --i) {
    #pragma unroll
    for (int j=0; j<nPCM; j++ ) {
      rows[j][i] = rows[j][i-1];
    }
  }
  rows[0][0] = frame_in[count];

  #pragma unroll nPCM
  for (int j=0; j<nPCM-1; j++) {
    //computation
    rows[j+1][0] = ...;
  }

  //computation of the final iteration
  frame_out[...] = ...;
 }
}
```

Listing 2 Stencil computation kernel

Listing 2 shows the stencil computation kernel written in OpenCL. The number of PEs in a PCM and the number of PCMs are denoted by *nPE* and *nPCM* respectively. Those are the two design parameters of the stencil computation architecture. Increasing *nPE* requires multiple computations done in parallel. As a result, more resources and more data are required. Since more data are required in a clock cycle, the required bandwidth is increased. Increasing *nPCM* increases the amount of resources.

3.3 Optimization Problem

We consider the following optimization problem to find the best accelerator architecture for stencil computation.

Objective function:

Minimization of the total number of clock cycles.

Constraints:

i. Resource utilization.
ii. Global memory bandwidth.

Freedom:

i. Number of PCMs (*nPCM*).
ii. Number of PEs per a PCM (*nPE*).

The objective function is computed according to Eq. (1). As shown in Listing 2, there are multiple loops in the stencil computation code. However, all the inner loops are completely unrolled, so that only the outer loop is considered to determine the number of cycles. Therefore, the number of clock cycles are "$II \times$ the number of loop iterations". Note that, to increase number of PEs, we manually unroll the outer loop by a factor of *nPE*. This reduces the number of clock cycles by a factor of *nPE* as shown in the first line of the stencil code. If the clock frequency is a constant, the number of clock cycles is relative to the processing time.

To compute the resource utilization, we consider all major resources of the FPGA, such as logic blocks, memory blocks, DSPs, etc. Different FPGAs contain different amount of resources. However, it may not be possible to achieve 100% utilization of all resources, due to placement difficulties on an FPGA. Moreover, routing become difficult for large designs and the clock frequency could drop considerably. Therefore, we often use less than 100% utilization of resources in order to predict performance accurately. For example, the resource utilization percentages for Stratix V

Table 1 Resource constraints of Stratix V FPGAs

Constraints
85% of the total logic modules
90% of the total registers
80% of the total memory blocks
100% of the total DSPs

FPGAs are shown in Table 1. It is possible to achieve this resource utilization without compromising on the clock frequency. According to our experience in many works such as [11, 12], large accelerators use around 80% of the FPGA resources. Therefore, we can say that such large accelerators can be implemented, while satisfying the resource constraint. For other FPGAs such as Arria 10, etc., resource constraints should be practically obtained.

The global memory access throughput of an FPGA board is limited by the theoretical bandwidth. The theoretical bandwidth is given by the FPGA board maker, or could be found in the profiling report of an OpenCL kernel. However, it is not possible to achieve 100% theoretical bandwidth. The achievable bandwidth depends on the efficiency of the memory controller, the memory access pattern, etc. For the stencil codes, we found that around 90% of the theoretical bandwidth can be achieved. The required bandwidth (B_{req}) of a kernel is given by Eq. (2). The size of the load and store data is the number of bytes accessed in cycle, and f is the clock frequency of the stencil computation accelerator. Since f is unknown at the design time, we use the maximum possible frequency of an FPGA board. This could be found by compiling a very small kernel code. Therefore, if $B_{req} \leq 0.9 \times B_{theory}$, we can say that the global memory bandwidth constraint is satisfied. Note that B_{theory} is the theoretical bandwidth.

$$B_{req} = (size[\text{load and store data}]) \times nPE \times f \tag{2}$$

Figure 5 shows the proposed automatic optimization methodology. It has two stages. In the first stage, performance of the stencil computation kernels with different design parameters are predicted. For each *nPE* value, we search for the optimum *nPCM* that provides the smallest number of clock cycles. This process is done automatically using a python-based program. According to the findings in [12], if *nPE* is a constant, the largest *nPCM* value provides the best performance. This is because, the clock frequency remains the same for different *nPCM* values. However, we cannot say which *nPE* value would provide the best accelerator architecture. This is because, slightly changing *nPE* can have a significant affect on *nPCM* and also the length of the pipeline. For example, if we reduce the *nPE* value by half, we can double the *nPCM* value, while maintaining the same degree of parallelism. According to our practical experience, offline compiler may not be able to maintain a high clock frequency for very large pipelines. Therefore, accelerators with different *nPE* could have different clock frequencies. Without knowing those clock frequencies, we may not be able to compare the performance of different accelerators.

In stage 2, we choose the parameter combinations that have different *nPE* values and the largest *nPCM*. Those are the candidate solutions, and perform the second stage compilation on those. After that, we can find the best solution by evaluation the processing time each one. Since the candidate solutions compiled in stage 2 is very few, we can decrease the design time significantly.

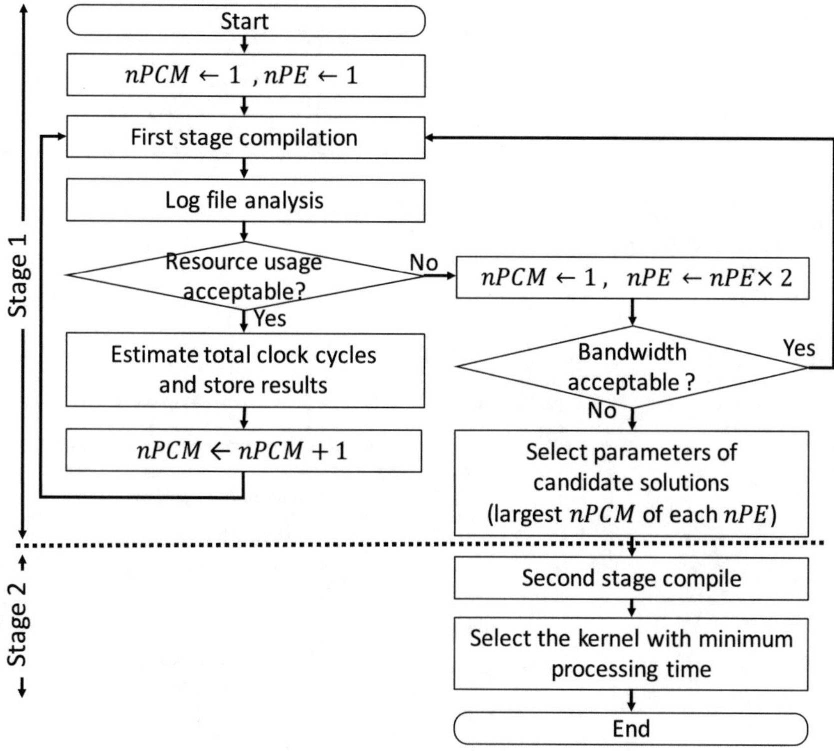

Fig. 5 The proposed automatic optimization methodology

4 Evaluation

For the evaluation, we use two FPGA boards. The resources of the FPGA boards are shown in the Table 2. FPGAs are configured using Intel FPGA SDK for OpenCL 16.1 [4]. We used the stencil computation applications shown in Table 3. Each example use a $4096 \times 32,768$ grid. Computations are done for 1,632 iterations.

Table 2 Resources of the FPGA boards used for the evaluation

Board name	DE5	395-D8
FPGA	5SGXEA7N2F45C2	5SGSED8N2F46C2LN
Logic modules	234,720	262,400
Registers	938,880	1,049,600
Internal memory	50.00 Mbits	50.13 Mbits
Memory blocks	2560	2567
DSP	256	1963
Theoretical bandwidth	25.6 GB/s	34.11 GB/s

Table 3 Stencil computation applications

Example	Computation
Laplace equation	$\left(V^t_{i,j-1} + V^t_{i-1,j} + V^t_{i+1,j} + V^t_{i,j+1}\right)/4$
2-D 5-point Jacobi	$c_1.V^t_{i,j-1} + c_2.V^t_{i-1,j} + c_3.V^t_{i,j} + c_4.V^t_{i+1,j} + c_5.V^t_{i,j+1}$

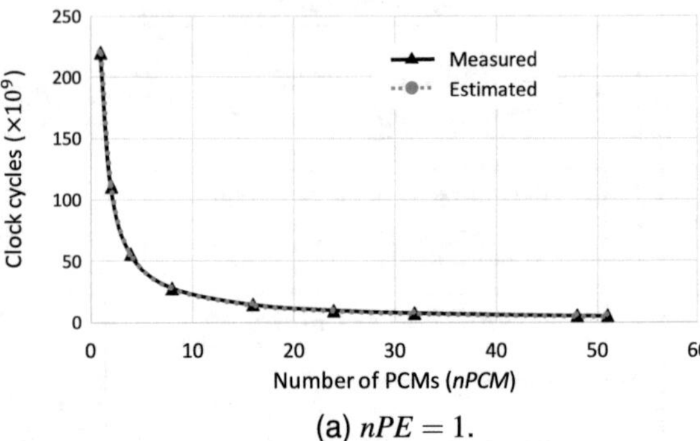

(a) $nPE = 1$.

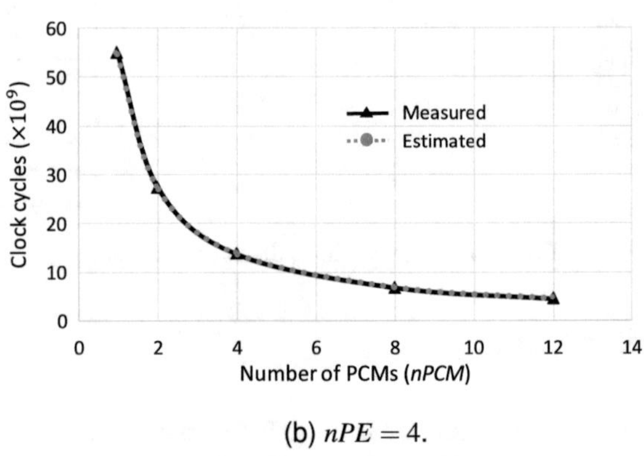

(b) $nPE = 4$.

Fig. 6 Estimated and measured clock cycles for 2-D 5-point stencil computation

4.1 Accuracy of the Processing Time Estimation

Figure 6 shows the relationship of the estimated clock cycles against the measured clock cycles for different parameter values. In Fig. 6a, $nPE = 1$ and $nPCM$ changes from 1 to 51. In Fig. 6b, $nPE = 4$ and $nPCM$ changes from 1 to 12. The estimated clock cycles are obtained using Eq. (1) as explained in Sect. 3. The measured clock cycles are obtained by multiplying the processing time by the clock frequency. According to the results, estimated results are very similar to the measured ones. Therefore, we can say that the estimation is very accurate.

Table 4 shows the estimated and measured clock cycles for different nPE values. The evaluation is done for 2-D 5-point stencil on DE5 board. In this evaluation, we increased $nPCM$ to the largest possible value that satisfy the constraints. According to the results, the number of estimated clock cycles are very similar for some solu-

Table 4 Estimated measured number of clock cycles for different *nPE* while increasing *nPCM* to the maximum

nPE	Estimated cycles	Measured cycles	Frequency (MHz)
1	4,301,655,232	4,302,639,000	297
2	4,432,567,524	4,434,804,934	298
4	4,565,077,184	4,569,681,156	304
8	4,564,241,600	4,629,408,652	295

Table 5 Comparison of the optimized architecture for different FPGA boards and stencil computation applications

Application	DE5 board		P395-D8 board	
	nPE	*nPCM*	*nPE*	*nPCM*
Laplace equation	8	17	8	18
2-D 5-point Jacobi	1	51	4	23

tions. Therefore, the actual clock frequency should be considered to find the optimal one. To find the clock frequency, we do the second stage compilation.

Table 5 shows the optimum parameters using different stencil computation examples and different FPGA boards. According to the results, we can see that the optimum parameters are quite different. As a result, the optimum accelerator architecture is also different. This shows that the optimization is necessary to achieve the full potential of an FPGA board for a given application.

4.2 Reduction of the Accelerator Design Time

Table 6 shows the design time of the proposed method and the conventional method explained in Sect. 2.1. In the proposed method, we consider the worst case design time by doing a complete search for all values of the design parameters in stage 1. Then we add it to the compilation time of stage 2 to get the design time. In the conventional method, we also consider the worst case design time by compiling all

Table 6 Comparison of the design time

Application	Design method	Time (hrs)
Laplace equation	Conventional	528.0
	Proposed	56.4
2-D 5-points Jacobi	Conventional	198.1
	Proposed	13.6

possible stencil computation accelerators. According to the results, the design time of the proposed method is 6–11% of that of the conventional method. We can reduce the design time further by replacing the complete search by more efficient search method, such as a binary search.

5 Conclusion

In this paper, we proposed an automatic optimization method for OpenCL kernels. We predict the performance of an OpenCL kernel by analyzing the log files and resource utilization reports. We use the proposed optimization methodology to implement stencil computation kernels. According to the evaluation, the performance prediction is very accurate, and the optimized design can be found in significantly smaller design time compared to the conventional method. In future, it could be possible to enhance the proposed method for other types of applications and also for NDRange kernels.

References

1. Czajkowski, T.S., Neto, D., Kinsner, M., Aydonat, U., Wong, J., Denisenko, D., Yiannacouras, P., Freeman, J., Singh, D.P., Brown, S.D.: OpenCL for FPGAs: Prototyping a Compiler, International Conference on Engineering of Reconfigurable Systems and Algorithms (ERSA), pp. 3–12 (2012)
2. Dohi, K., Okina, K., Soejima, R., Shibata, Y., Oguri, K.: Performance modeling of stencil computing on a stream-based FPGA accelerator for efficient design space exploration. IEICE Trans. Inf. Syst. **E98-D**(2), 298–308 (2015)
3. Khronos Group: The open standard for parallel programming of heterogeneous systems. https://www.khronos.org/opencl/
4. Intel: SDK for OpenCL. https://www.altera.com/products/design-software/embedded-software-developers/opencl/overview.html
5. Jia, Q., Zhou, H.: Tuning stencil codes in OpenCL for FPGAs. In: IEEE 34th International Conference on Computer Design (ICCD), pp. 249–256 (2016)
6. Krommydas, K., Sasanka, R., Feng, W.: GLAF: a visual programming and auto-tuning framework for parallel computing. In: IEEE 44th International Conference on Parallel Processing (ICPP), pp. 859–868 (2015)
7. Krommydas, K., Sasanka, R., Feng, W.: Bridging the FPGA programmability-portability Gap via automatic OpenCL code generation and tuning. In: IEEE 27th International Conference on Application-Specific Systems, Architectures and Processors (ASAP), pp. 213–218 (2016)
8. Karniadakis, G., Sherwin, S.: Spectral/hp Element Methods for Computational Fluid Dynamics. Oxford University Press (2013)
9. Roth, G., Mellor-Crummey, J., Kennedy, K., Brickner, R.G.: Compiling stencils in high performance fortran. In: Proceedings of the 1997 ACM/IEEE Conference on Supercomputing, pp. 1–20 (1997)
10. Sano, K., Hatsuda, Y., Yamamoto, S.: Multi-FPGA accelerator for scalable stencil computation with constant memory bandwidth. IEEE Trans. Parallel Distrib. Syst. **25**(3), 695–705 (2014)
11. Waidyasooriya, H.M., Hariyama, M.: Hardware-Acceleration of short-read alignment based on the burrows-wheeler transform. IEEE Trans. Parallel Distrib. Syst. **27**(5), 1358–1372 (2016)

12. Waidyasooriya, H.M., Takei, Y., Tatsumi, S., Hariyama, M.: OpenCL-Based FPGA-Platform for stencil computation and its optimization methodology. IEEE Trans. Parallel Distrib. Syst. **28**(5), 1390–1402 (2017)
13. Yee, K.S.: Numerical solution of initial boundary value problems involving Maxwells equations in isotropic media. IEEE Trans. Antennas Propag. **14**(3), 302–307 (1966)

Should Duration and Team Size Be Used for Effort Estimation?

Takeshi Kakimoto, Masateru Tsunoda and Akito Monden

Abstract Project management activities such as scheduling and project progress management are important to avoid project failure. As a basis of project management, effort estimation plays a fundamental role. To estimate software development effort by mathematical models, variables which are fixed before the estimation are used as independent variables. Some studies used team size and project duration as independent variables. Although they are sometimes fixed because of the limitation of human resources or business schedule, they may change by the end of the project. For instance, when delivery is delayed, actual duration and estimated duration is different. So, although using team size and project duration may enhance estimation accuracy, the error may also lower the accuracy. To help practitioners to select independent variables, we analyzed whether team size and duration should be used or not, when we consider the error included in the team size and the duration. In the experiment, we assumed that duration and team size include errors when effort is estimated. To analyze influence of the errors, we add n% errors to duration and team size. As a result, using duration as an independent variable was not very effective in many cases. In contrast, using maximum team size as an independent variable was effective when the error rate is equal or less than 50%.

Keywords Software effort prediction · Project management · Productivity · Estimation error

T. Kakimoto
Department of Electrical and Computer Engineering,
National Institute of Technology, Kagawa College, Takamatsu, Japan
e-mail: kakimoto@t.kagawa-nct.ac.jp

M. Tsunoda (✉)
Department of Informatics, Kindai University, Higashiosaka, Japan
e-mail: tsunoda@info.kindai.ac.jp

A. Monden
Graduate School of Natural Science and Technology, Okayama University,
Okayama, Japan
e-mail: monden@okayama-u.ac.jp

R. Lee (ed.), *Software Engineering, Artificial Intelligence, Networking and Parallel/Distributed Computing*, Studies in Computational Intelligence 721,
DOI 10.1007/978-3-319-62048-0_7

1 Introduction

As recent software systems grow in size and complexity, project management activities such as staffing, scheduling and project progress management are becoming increasingly important to avoid project failure (cost overrun and/or delayed delivery). As a basis of project management, effort estimation plays a fundamental role; therefore, accurate effort estimation is vital to organization's profitability.

To date, various estimation models that use past projects' historical data have been proposed [2, 23, 23]. One of the most commonly used estimation models is a linear regression model, which represents the relationship between the dependent variable (i.e. effort) and independent variables such as functional size, architecture, programming language, and so on.

Analogy based estimation [23] is one of major estimation methods, and many proposals and case studies have been reported [8, 9, 20, 25, 29]. Analogy based estimation selects projects (neighborhood projects) which are similar to the estimated project from past project dataset, and estimates effort based on similar projects' effort. One of the advantages of analogy based estimation is that estimation results are comprehensible for estimators such as project managers [29], because they can confirm neighborhood projects used for estimation.

To estimate software development effort by mathematical models, variables which are fixed before the estimation are used as independent variables. Effort is estimated on the early phase of projects, i.e., after basic design phase. For example, architecture and programming language are fixed after the phase, and they are often used as independent variables of estimation models. In contrast, variables which are not fixed before the estimation cannot be used as the independent variables.

Some studies used team size [7, 16] and project duration [1, 12, 13] as independent variables. However, they are not always fixed before the estimation. They are occasionally fixed after estimation. For example, when estimated effort is 9 person-months, duration is set as 9 and team size is set as 3. Sometimes, they are fixed because of the limitation of human resources or business schedule. But they may change by the end of the project. For instance, when delivery is delayed, actual duration (fixed after project is finished) and estimated duration (value input to the model) is different. Generally, estimation model is made based on the actual duration of past projects, and therefore input value (estimated effort) would include errors, as shown in Fig. 1. Therefore, although using team size and project duration may enhance estimation accuracy, the error may also lower the accuracy.

The goal our study is to help practitioners to select independent variables when they build effort estimation models. So, we analyzed whether team size and duration should be used or not, when we consider the error included in the team size and the duration. To clarify the purpose of the analysis, we set following research questions:

- **RQ1**: Is duration effective to improve estimation accuracy?
- **RQ2**: Is team size effective to improve estimation accuracy?
- **RQ3**: At what error rate is estimation accuracy negatively affected?

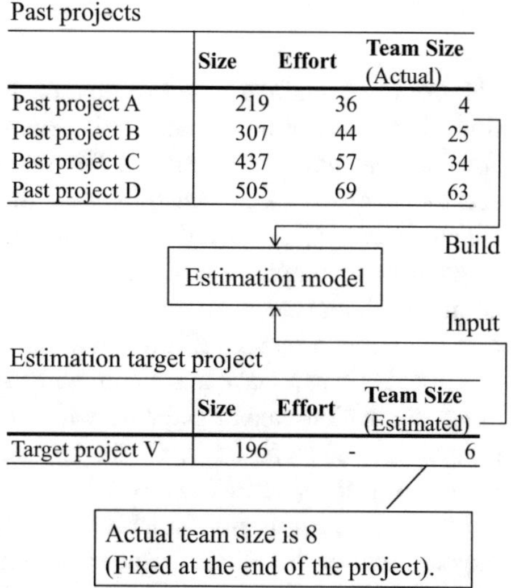

Past projects
Size
Effort
Team Size (Actual)
Past project A 219 36 4
Past project B 307 44 25
Past project C 437 57 34
Past project D 505 69 63
Build
Estimation model
Input
Estimation target project
Size
Effort
Team Size (Estimated)
Target project V 196 - 6
Actual team size is 8
(Fixed at the end of the project).

Fig. 1 An example of the error included in team size

Section 2 explains effort estimation methods used in the experiment. Section 3 describes the experimental setting, and Sect. 4 shows results of the experiment. Section 5 explains related work, and Sect. 6 concludes the paper.

2 Effort Estimation Methods

2.1 Multiple Linear Regression Model

The multiple linear regression model is widely used when estimating software development effort mathematically. The model is built based on ordinary least squares. When the effort is denoted as y, and independent variables such as software size are denoted as $x_1, x_2, \ldots, x_k$ (k is the number of independent variables), y is explained as:

$$y = \beta_0 + \beta_1 x_1 + \beta_2 x_2 + \cdots + \beta_k x_k + \varepsilon \quad (1)$$

In the equation, β_0 is an intercept, $\beta_1, \beta_2, \ldots, \beta_k$ are partial regression coefficients, and ε is an error term. As a rule of thumb, to build a proper model using linear regression analysis, it is needed that the number of data points is five to ten times larger than the number of independent variables.

When building the model, log-transformation is applied to enhance the accuracy of the model [10]. This is because the distributions of some variables such as effort and software size are log-normal distribution.

2.2 Analogy Based Estimation

The origin of analogy based estimation is CBR (case based reasoning), which is studied in artificial intelligence field. Shepperd et al. [22] applied CBR to software development effort estimation. CBR selects a case similar to current issue from accumulated past cases, and applies solution of the case to the issue. CBR assumes similar issues can be solved by similar solution. Analogy based estimation assumes neighborhood (similar) projects (For example, development size and used programming language is similar) have similar effort, and estimates effort based on neighborhood projects' effort. Although ready-made estimation models such as COCOMO [2] can make estimation without stored software project dataset, analogy based estimation cannot estimate without it. It is a weak point of analogy based estimation, but it can be overcome by using public dataset.

Analogy based estimation uses $k \times l$ matrix shown in Table 1. In the matrix, p_i is i-th project, m_{ij} is j-th variable. That is, each row denotes a data point (i.e., a project), and each columns denotes a metric. We presume p_a is estimation target project, and $\hat{m}_{ab}$ is the estimated value of m_{ab}. Procedures of analogy based estimation consist of the three steps described below.

Step 1: Since each variable has different range of value, this step makes the ranges [0, 1]. The value m'_{ij}, normalized the value of m_{ij} is calculated by:

$$m'_{ij} = \frac{m_{ij} - min(m_j)}{max(m_j) - min(m_j)} \tag{2}$$

In the equation, max(m_j) and min(m_j) denote the maximum and minimum value of m_j respectively. The equation is one of the commonly used methods to normalize the range of a value [24].

Step 2: To find projects which are similar to estimated project p_a (i.e., identifying neighborhood projects), similarity between p_a and other projects p_i is calculated. Variables of p_a and p_i are used as elements of vectors, and cosine of the vectors are regarded as similarity. Similarity $sim(p_a, p_i)$ between p_a and p_i is calculated by:

Table 1 Dataset used on analogy based effort estimation

	*Variable*1	*Variable*2	…	*Variablej*	…	*Variablel*
p_1	m_{11}	m_{12}	…	m_{1j}	…	m_{1l}
p_2	m_{21}	m_{22}	…	m_{2j}	…	m_{2l}
…	…	…		…		…
p_i	m_{i1}	m_{i2}	…	m_{ij}	…	m_{il}
…	…	…		…		…
p_k	m_{k1}	m_{k2}	…	m_{kj}	…	m_{kl}

$$sim(p_a, p_i) = \frac{\sum_{j \in M_a \cap M_i} \left(m'_{aj} - avg\left(m'_j\right)\right)\left(m'_{ij} - avg\left(m'_j\right)\right)}{\sqrt{\sum_{j \in M_a \cap M_i} \left(m'_{aj} - avg\left(m'_j\right)\right)}\sqrt{\sum_{j \in M_a \cap M_i} \left(m'_{ij} - avg\left(m'_j\right)\right)}} \quad (3)$$

In the equation, M_a and M_i are set of variables measured in project p_a and p_i respectively. $avg(m', j)$ is average of i-th variable. The range of $sim(p_a, p_i)$ is [−1, 1].

Step 3: The estimated effort of project p_a is calculated by actual effort of k neighborhood projects. While average of neighborhood projects' effort is generally used, we adopt size adjustment method, which showed high estimation accuracy in some studies [9, 20, 29]. Estimated value $\hat{m}ab$ is calculated by:

$$\hat{m}_{ab} = \frac{\sum_{i \in k-nearest\ Projects} (m_{ib} \times amp(p_a, p_i) \times sim(p_a, p_i))}{\sum_{i \in k-nearest\ Projects} sim(p_a, p_i)} \quad (4)$$

$$amp(p_a, p_i) = \frac{fp_a}{fp_i} \quad (5)$$

In the equation, fp_a and fp_i are software size of project p_a and p_i respectively. Size adjustment method assumes effort is s times (s is real number greater than 0) larger when software size is s times larger. The method adjusts effort of p_i based on ratio of target project's size fp_a and neighborhood project's size fp_i.

3 Experiment

3.1 Datasets

We used the ISBSG [6], Kitchenham [11], and Desharnais datasets [5]. Nominal scale variables were transformed into dummy variables (e.g. if the variable has n categories, it is transformed into $n - 1$ dummy variables). We removed dummy variables when the number of cases which correspond with the category was very small.

The ISBSG dataset is provided by the International Software Benchmark Standard Group (ISBSG), and it includes project data collected from software development companies in 20 countries [6]. The dataset (Release 9) includes 3026 projects that were carried out between 1989 and 2004, and 99 variables were recorded. The ISBSG dataset includes low quality project data (Data quality ratings are also included in the dataset).

We extracted projects based on the previous study [15] (Data quality rating is A or B, function point was recorded by the IFPUG method, and so on). Also, we excluded projects that included missing values (listwise deletion). As a result, we used 196 projects. The variables used in our experiment are shown in Table 2. They are almost same as the previous study [15] except for duration and maximum team size.

The Kitchenham dataset includes 145 projects of a software development company, shown by Kitchenham et al. in their study [11]. We selected 135 projects that do not include missing values. Three variables shown in Table 3 were chosen as the independent variables, and inadequate variables for effort estimation (e.g. estimated effort by a project manager) were eliminated. Development type was transformed into dummy variables.

The Desharnais dataset includes 88 projects of 1980s, collected from a Canadian company by Desharnais [5]. The dataset is available at the PROMISE Repository [3]. We used 77 projects that do not have missing values. Variables shown in Table 4 were used as independent variables, and development year were not used. Also, the

Table 2 Variables of ISBSG dataset

Variable	Scale	Description
FP	Ratio	Unadjusted function point
Effort	Ratio	Summary work effort (hour)
Duration	Ratio	Actual duration of project
Maximum team size	Ratio	Maximum number of personnel who engaged the project
Language type	Ratio	3GL (second-generation programming language), 4GL, and others
Development type	Nominal	New development, enhancement, and others
Development platform	Nominal	Mid range, main frame, and others

Table 3 Variables of Kitchenham dataset

Variable	Scale	Description
FP	Ratio	Adjusted function point
Effort	Ratio	Actual development effort (hour)
Duration	Ratio	Actual duration of project
Development type	Nominal	Development, perfective, and others

Table 4 Variables of Desharnais dataset

Variable	Scale	Description
FP	Ratio	Unadjusted function point
Effort	Ratio	Actual development effort (hour)
Duration	Ratio	Actual duration of project
Adjustment factor	Ratio	Adjustment factor of function point
TeamExp	Interval	Experience of team (measured in years)
ManagerExp	Interval	Experience of manager (measured in years)
Language	Nominal	Type1, Type2, and others

adjusted function point, the number of transactions, and the number of entities were not used to avoid multicollinearity. Programming language was transformed into dummy variables which reflects different development environments.

3.2 Evaluation Criteria

To evaluate the accuracy of effort estimation, we used the conventional metrics such as *AE* (Absolute Error), *MRE* (Magnitude of Relative Error) [4], and *BRE* (Balanced Relative Error) [21]. Especially, *MRE* is widely used to evaluate the effort estimation accuracy [29].

When x denotes actual effort, and $\hat{x}$ denotes estimated effort, each criterion is calculated by the following equations:

$$AE = |x - \hat{x}| \tag{6}$$

$$MRE = \frac{|x - \hat{x}|}{x} \tag{7}$$

$$BRE = \left\{ \begin{array}{l} \frac{|x - \hat{x}|}{x}, x - \hat{x} \geq 0 \\ \frac{|x - \hat{x}|}{\hat{x}}, x - \hat{x} < 0 \end{array} \right\} \tag{8}$$

A lower value of each criterion indicates higher estimation accuracy. Intuitively, *MRE* means error relative to actual effort. However, *MRE* have biases for evaluating under estimation [14]. The maximum *MRE* is 1 even if an extreme underestimate occurs (For instance, when the actual effort is 1000 person-hour, and the estimated effort is 0 person-hour, *MRE* is 1). So we employed *BRE* whose evaluation is not biased as is both *MRE* [22], and we evaluated the classified models based on mainly *BRE* (*MRE* were used for reference). We did not use Pred(25) [4] which is sometimes used as an evaluation criterion, because Pred(25) is based on *MRE* and it has also a bias for evaluating under estimation.

3.3 Procedure of Experiment

In the experiment, we assume that duration and maximum team size include errors when effort is estimated. This is because they are not fixed when effort is estimated (i.e., they are estimated values). At the end of the project, actual values of them may be different from the estimated values. To analyze influence of the errors, we add n % errors to duration and maximum team size. We set n as 0, 25, 50, 100, and 200%. The definition of the error rate is same as *BRE*.

Figure 2 is an example of the procedure. Dataset is divided into learning data and test data. Only team size on the test data includes the error. In the figure, team size on test data includes 25% errors. We generated new values of team size including the errors, and used it when effort is estimated.

We made the following models in the experiment, using analogy based estimation and multiple linear regression analysis.

(A) Models without duration and maximum team size
(B) Models with duration
(C) Models with maximum team size
(D) Models with duration and maximum team size

On model A, independent variables do not include duration and maximum team size. Model B includes duration as one of independent variables. In the same way, model C and D have independent variables. We call the model A as baseline, and evaluated other models with the baseline. Model C and D were made when ISBSG dataset is used. Since only ISBSG dataset includes maximum team size.

We evaluated accuracies of models by differences of criteria from a baseline model. Therefore, positive values mean estimation accuracies were improved from the baseline model, and negative values mean estimation accuracies got worse. We applied fivefold cross validation to divide the dataset into fit datasets and test datasets. The fit datasets were used to build the models, and the test datasets were used to evaluate the models.

Fig. 2 Injecting errors into values of an independent variable

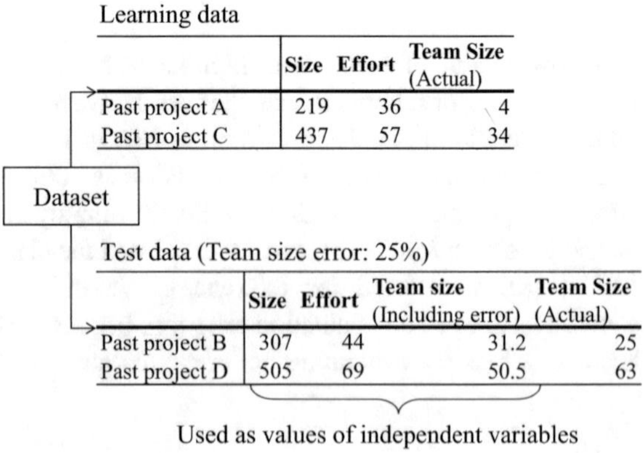

Dataset

Learning data

	Size	Effort	Team Size (Actual)
Past project A	219	36	4
Past project C	437	57	34

Test data (Team size error: 25%)

	Size	Effort	Team size (Including error)	Team Size (Actual)
Past project B	307	44	31.2	25
Past project D	505	69	50.5	63

Used as values of independent variables

Logarithmic transformation and variable selection was applied when multiple regression models were built. The number of neighborhoods was set as 5 when analogy based estimation was applied.

4 Results

4.1 Preliminary Analysis

As preliminary analysis, we analyzed the relationship of duration and team size to effort and productivity. If the relationship is strong, using duration and team size as independent variables is expected to enhance estimation accuracy. Productivity was calculated by FP (function point) divided by effort. Strength of the relationship was evaluated using Spearman's rank correlation coefficient.

The result is shown in Table 5. The relationship between duration and productivity was weak on the three datasets, although the relationship between duration and effort was not weak. The result suggests that using duration as an independent variable is not very effective to enhance estimation accuracy. In contrast, strength of the relationship between maximum team size and productivity was moderate on ISBSG dataset. So, using maximum team size as an independent variable is expected to enhance estimation accuracy.

4.2 Estimation Accuracy of Analogy Based Estimation

Table 6 shows estimation accuracy of the models when analogy based estimation was used. In the table, top row of each dataset shows the accuracy of the model A (i.e., the baseline), and other rows do the difference from the baseline. Boldface in the table indicates the accuracy is improved using duration and maximum team size as independent variables.

Evaluation of model B (using duration): On Desharnais dataset, even the error rate is 0%, improvement of the accuracy was very small. Specifically, improvement of average *AE* and average *BRE* were very small, and median *AE* and median *BRE* got slightly worse. When the error rate of duration was equal or less than 50%, the negative influence to estimation accuracy was small. On Kitchenham dataset,

Table 5 Relationship to effort and productivity

Dataset	Variable	Effort	Productivity
Desharnais	Duration	0.57	–0.14
Kitchenham	Duration	0.57	–0.23
ISBSG	Duration	0.59	–0.17
ISBSG	Max. team size	0.68	–0.47

Table 6 Estimation accuracy of models when using analogy based estimation

Dataset	Variables	Error rate (%)	Average *AE*	Median *AE*	Average *MRE* (%)	Median *MRE* (%)	Average *BRE* (%)	Median *BRE* (%)
Desharnais	Duration (Model B)	–	1884	1880	50	49	56	46
		0	−38	**280**	−1	**1**	0	0
		25	−9	**249**	−1	**1**	0	**3**
		50	−72	**50**	−2	**0**	−1	**2**
		100	−192	−221	−6	−5	−5	−2
		200	−573	−681	−20	−26	−19	−18
Kitchenham	Duration (Model B)	–	1630	3221	79	116	100	61
		0	−46	−566	−2	−9	**6**	**13**
		25	−105	−673	−6	−26	**2**	**11**
		50	−96	−732	0	−6	**6**	**10**
		100	−403	−959	−22	−25	−17	−10
		200	−813	−1384	−47	−58	−48	−26
ISBSG	Duration (Model B)	–	4444	6345	155	247	188	104
		0	−781	−3519	−76	−346	−68	**5**
		25	−463	−1346	−77	−324	−71	−8
		50	−516	−1004	−69	−251	−64	−6
		100	−579	−637	−78	−244	−74	−6
		200	−561	−156	−76	−201	−74	−22
ISBSG	Team size (Model C)	–	4444	6345	155	247	188	104
		0	**946**	**1346**	**9**	−60	**22**	**26**
		25	**925**	**1356**	**4**	−81	**18**	**26**
		50	**793**	**1183**	−7	−98	**6**	**27**
		100	**425**	**883**	−24	−142	−13	**12**
		200	−444	**202**	−78	−241	−71	−29
ISBSG	Duration Team size (Model D)	–	4444	6345	155	247	188	104
		0	**426**	−502	−38	−277	−19	**37**
		25	**209**	−371	−50	−271	−33	**28**
		50	−88	−413	−63	−285	−51	**19**
		100	−678	−753	−86	−326	−78	−11
		200	−1919	−1879	−169	−486	−176	−56

average and median *BRE* were improved when the error rate was less than 100%. However, average and median *AE* got worse even the rate was 0%. On ISBSG dataset, estimation accuracy got worse on most cases. Therefore, when effort is estimated by analogy based estimation, using duration as an independent variable is not effective but sometimes negatively affects to estimation accuracy.

Evaluation of model C and D (using maximum team size): When maximum team size was used as an independent variable on ISBSG dataset (model C), it was effective to improve estimation accuracy. Except for median *MRE*, estimation accuracy was improved on most cases, when the error rate was equal or less than

50%. When both maximum team size and duration were used (model D), median *AE* and average *BRE* got worse. This would be because duration was negatively affected to the accuracy. So, using maximum team size as an independent variable is effective when the error rate is equal or less than 50%, and effort is estimated by analogy based estimation.

Table 7 Estimation accuracy of models when using multiple linear regression analysis

Dataset	Variables	Error rate (%)	Average *AE*	Median *AE*	Average *MRE* (%)	Median *MRE* (%)	Average *BRE* (%)	Median *BRE* (%)
Desharnais	Duration (Model B)	–	1652	1014	36	29	47	35
		0	**7**	−56	**1**	−2	**1**	−3
		25	**37**	−75	**1**	−4	**1**	−4
		50	**22**	−59	**1**	−5	**1**	−6
		100	−84	−166	−1	−4	−2	−10
		200	−301	−370	−6	−5	−9	−5
Kitchenham	Duration (Model B)	–	1806	627	70	40	93	52
		0	**203**	**18**	**5**	**4**	**12**	**3**
		25	**185**	**38**	**6**	**2**	**12**	**5**
		50	**138**	**76**	**5**	**1**	**9**	**2**
		100	**20**	−51	−1	−4	−2	−11
		200	−255	−247	−16	−19	−29	−43
ISBSG	Duration (Model B)	–	3496	1690	101	55	148	92
		0	**3**	**133**	**4**	−1	0	**6**
		25	**8**	**150**	**3**	0	−2	0
		50	−3	**122**	**3**	**2**	−4	**3**
		100	−33	**140**	**1**	**2**	−10	**5**
		200	−143	**230**	−5	0	−21	**1**
ISBSG	Team size (Model C)	–	3496	1690	101	55	148	92
		0	**636**	**570**	**12**	**13**	**31**	**33**
		25	**434**	**494**	**5**	**13**	**22**	**37**
		50	**134**	**322**	−5	**6**	**8**	**27**
		100	−556	−57	−29	−2	−27	**8**
		200	−1911	−735	−79	−16	−104	−34
ISBSG	Duration Team size (Model D)	–	3496	1690	101	55	148	92
		0	**929**	**691**	**21**	**20**	**40**	**46**
		25	**583**	**648**	**9**	**14**	**23**	**36**
		50	−28	**324**	−9	**5**	−6	**19**
		100	−1354	−386	−54	−10	−77	−25
		200	−4030	−1993	−151	−27	−238	−111

4.3 Estimation Accuracy of Multiple Regression Analysis

Table 7 shows estimation accuracy of the models when multiple regression analysis was used. The structure of the table is same as Table 6.

Evaluation of model B (using duration): On Desharnais dataset, average *AE*, *MRE*, and *BRE* were slightly improved, when the error rate was equal or less than 50%. In contrast, median *AE*, *MRE*, and *BRE* got worse. On Kitchenham dataset, estimation accuracy was improved when the error rate was equal or less than 50%. Especially, the improvement of average *BRE* was about 10%. On ISBSG dataset, using duration did not affect estimation accuracy very much, when the error rate was equal or less than 50%. Overall, using duration as independent variable did not negatively affected when multiple regression analysis was used, and sometimes positively affected when the error rate was equal or smaller than 50%.

Evaluation of model C and D (using maximum team size): When maximum team size was used as an independent variable on ISBSG dataset (model C), it was effective to improve estimation accuracy. Estimation accuracy was improved on most cases, when the error rate was equal or less than 50%. Also, when both maximum team size and duration were used (model D), estimation accuracy was improved when the error rate is equal or less than 25%. When the error rate was 0%, the estimation accuracy of model D was better than the model C. However, then the rate is 25%, the accuracy was almost same. Therefore, using maximum team size as an independent variable is effective when the error rate is equal or less than 50%, but adding duration as an independent variable does not improve estimation accuracy unless the error rate is very small.

4.4 Summery of the Results

Using duration as an independent variable (model B) was not very effective in many cases. Estimation accuracy was explicitly improved only when multiple regression analysis was used on Kitchenham dataset. So, the answer of RQ1 is "No."

In contrast, using maximum team size as an independent variable (model C) was effective when the error rate is not very large (equal or less than 50%). So, the answer of RQ2 is "Yes." To know the error rate, duration and maximum team size should be estimated and recorded, and we can calculate the rate when the data is accumulated.

When the error rate is equal or more than 100%, the estimation accuracy got worse in many cases. So, the answer of RQ3 is "100% and more." Overall, influence of duration, maximum team size and the error rate to estimation accuracy was not very different between analogy based estimation and multiple regression analysis. So, the influence would not be very different even when other estimation models are used.

5 Related Work

In our past studies, we focus on error included in independent variables such as difference between estimated team size and actual team size. Study [27] proposed an estimation method based on stratification of team size, and analyzed the influence of the error of team size. Also, study [26] proposed an estimation method based on productivity and proposed new method to absorb the influence of the error of the estimated productivity. However, study [27] used team size as a categorical variable, and not used as a ratio scale variable. Also, study [26] used productivity, but not used team size as an independent variable. Therefore, our past studies [26, 27] did not clarify the effect of team size and duration to estimation accuracy.

There are many studies which analyzed the relationship between project attributes such as duration and productivity. For example, Maxwell et al. [17] and Premraj et al. [22] analyzed an influence of business sector for productivity, using Finnish software development project dataset collected by Software Technology Transfer Finland (STTF). Lokan et al. [16] showed productivity by business sector using dataset of International Software Benchmarking Standards Group (ISBSG). In these studies, projects for manufacturing have the highest productivity, and projects for banking/Insurance have the lowest productivity.

Also, relationship of team size and duration to productivity was analyzed in some studies [18, 28]. In the study [28], team size showed strong relationship to productivity, and duration was weak relationship to productivity. Dataset used in the study is Japanese cross-company dataset, and it is not ISBSG dataset. Therefore, our analysis result has external validity to some extent.

6 Conclusions

In this study, we evaluated the effect of using project duration and maximum team size as an independent variable on effort estimation models. We assume that duration and maximum team size include errors when effort is estimated. This is because they are not fixed on the point. To analyze influence of the errors, we add n % errors to duration and maximum team size. We set n as 0, 25, 50, 100, and 200%. We used ISBSG dataset, Kitchenham dataset, and Desharnais datasets in the experiment. To estimate effort, analogy based estimation and multiple linear regression analysis were used. Our findings include the followings:

- Using duration as an independent variable was not very effective in many cases.
- Using maximum team size as an independent variable was effective when the error rate is not very large (equal or less than 50%).
- When the error rate is equal or more than 100%, the estimation accuracy got worse in many cases.

- Influence of duration, maximum team size, and the error rate to estimation accuracy was not very different between analogy based estimation and multiple regression analysis.

The influence of maximum team size was evaluated only one dataset. To enhance the reliability of the results, we will analyze the influence in other dataset.

Acknowledgements This research was partially supported by the Japan Ministry of Education, Science, Sports, and Culture [Grant-in-Aid for Scientific Research (C) (No. 16K00113)].

References

1. Azzeh, M., Neagu, D., Cowling, P.: Fuzzy grey relational analysis for software effort estimation. Empir. Softw. Eng. **15**(1), 60–90 (2010)
2. Boehm, B.: Software engineering economics. Prentice Hall (1981)
3. Boetticher, G. Menzies, T., Ostrand, T.: PROMISE repository of empirical software engineering data. West Virginia University, Department of Computer Science (2007)
4. Conte, S., Dunsmore, H., Shen, V.: Software Engineering, Metrics and Models. Benjamin/Cummings, Redwood City (1986)
5. Desharnais, J.: Analyse Statistique de la Productivitie des Projets Informatique a Partie de la Technique des Point des Function. Master thesis, University of Montreal, 1989 (1989)
6. International Software Benchmarking Standards Group (ISBSG): ISBSG Estimating: Benchmarking and research suite, ISBSG (2004)
7. Jeffery, R., Ruhe, M., Wieczorek, I.: Using public domain metrics to estimate software development effort. In: Proceedings of the International Symposium on Software (METRICS), pp. 16–27 (2001)
8. Keung, J., Kitchenham, B., Jeffery, R.: Analogy-X: providing statistical inference to analogy-based software cost estimation. IEEE. Trans. Softw. Eng. **34**(4), 471–484 (2008)
9. Kirsopp, C., Mendes, E., Premraj, R., Shepperd, M.: An empirical analysis of linear adaptation techniques for case-based prediction. In: Proceedings of International Conference on Case-Based Reasoning, pp. 231–245 (2003)
10. Kitchenham, B., Mendes, E.: Why comparative effort prediction studies may be invalid. In: Proceedings of International Conference on Predictor Models in Software Engineering (PROMISE), art 4, p. 5 (2009)
11. Kitchenham, B., Pfleeger, S., McColl, B., Eagan, S.: An empirical study of maintenance and development estimation accuracy. J. Syst. Softw. **64**(1), 57–77 (2004)
12. Li, Y., Xie, M., Goh, T.: A study of the non-linear adjustment for analogy based software cost estimation. Empir. Softw. Eng. **14**(6), 603–643 (2009)
13. Li, J., Ruhe, G.: Analysis of attribute weighting heuristics for analogy-based software effort estimation method AQUA+. Empir. Softw. Eng. **13**(1), 63–96 (2008)
14. Lokan, C.: What should you optimize when building an estimation model? In: Proceedings of International Software Metrics Symposium (METRICS), p. 34. Como, Italy (2005)
15. Lokan, C., Mendes, E.: Cross-company and single-company effort models using the ISBSG Database: a further replicated study. In: Proceedings of the International Symposium on Empirical Software Engineering (ISESE), pp. 75–84 (2006)
16. Lokan, C., Wright, T., Hill, P., Stringer, M.: Organizational benchmarking using the ISBSG data repository. IEEE Softw. **18**(5), 26–32 (2001)
17. Maxwell, K., Forselius, P.: Benchmarking software development productivity. IEEE Softw. **17**(1), 80–88 (2000)

18. Maxwell, K., Wassenhove, L., Dutta, S.: Software development productivity of european space, military, and industrial applications. IEEE Trans. Softw. Eng. **22**(10), 706–718 (1996)
19. Mendes, E., Mosley, N., Counsell, S.: A replicated assessment of the use of adaptation rules to improve web cost estimation. In: Proceedings of the International Symposium on Empirical Software Engineering (ISESE), pp. 100–109 (2003)
20. Miyazaki, Y., Terakado, M., Ozaki, K., Nozaki, H.: Robust regression for developing software estimation models. J. Syst. Softw. **27**(1), 3–16 (1994)
21. Mølokken-Østvold, K., Jørgensen, M.: A comparison of software project overruns-flexible versus sequential development models. IEEE Trans. Softw. Eng. **31**(9), 754–766 (2005)
22. Premraj, R., Shepperd, M., Kitchenham, B., Forselius, P.: An empirical analysis of software productivity over time. In: Proceedings of International Software Metrics Symposium (METRICS), p. 37 (2005)
23. Srinivasan, K., Fisher, D.: Machine learning approaches to estimating software development effort. IEEE Trans. Softw. Eng. **21**(2), 126–137 (1995)
24. Strike, K., Eman, K., Madhavji, N.: Software cost estimation with incomplete data. IEEE Trans. Softw. Eng. **27**(10), 890–908 (2001)
25. Tosun, A., Turhan, B., Bener, A.: Feature weighting heuristics for analogy-based effort estimation models. Expert. Syst. Appl. **36**(7), 10325–10333 (2009)
26. Tsunoda, M., Monden, A., Keung, J., Matsumoto, K.: Incorporating expert judgment into regression models of software effort estimation. In: Proceedings of Asia-Pacific Software Engineering Conference (APSEC), pp. 374–379 (2012)
27. Tsunoda, M., Monden, A., Matsumoto, K., Takahashi, A.: Software development effort estimation models stratified by productivity factors. SEC J (in Japanese) 58–67 (2009)
28. Tsunoda, M., Monden, A., Yadohisa, H., Kikuchi, N., Matsumoto, K.: Software development productivity of Japanese enterprise applications. Inf. Technol. Manage. **10**(4), 193–205 (2009)
29. Walkerden, F., Jeffery, R.: An empirical study of analogy-based software effort estimation. Empir. Softw. Eng. **4**(2), 135–158 (1999)

A Lifelog Data Portfolio for Privacy Protection Based on Dynamic Data Attributes in a Lifelog Service

Prajak Chertchom, Shigeaki Tanimoto, Hayato Ohba, Tsutomu Kohnosu, Toru Kobayashi, Hiroyuki Sato and Atsushi Kanai

Abstract With the challenge of contemporary mobile computing, quantified self-data need a well-established data service model based on information sources which takes individual privacy into account. This paper presents a lifelog attribute data portfolio (LLADP) that will be used for practically modeling life events and for digitizing such information. In this article, we also propose the privacy implications of lifelogging for each attribute. We designed an attribute portfolio on the basis of the kinds of lifelog services that are already provided in contemporary wearable devices. We aim to map real life models with computerized data models. However, life events may be impossible to completely record because of current device limitations. Thus, we aim to propose a lifelog attribute data portfolio (LLADP) using hybrid cloud management that takes privacy implications and a basic privacy policy into account.

P. Chertchom
Thai-Nichi Institute of Technology, Bangkok, Thailand
e-mail: prajak@tni.ac.th

S. Tanimoto (✉) · H. Ohba · T. Kohnosu
Chiba Institute of Technology, Narashino, Japan
e-mail: shigeaki.tanimoto@it-chiba.ac.jp

H. Ohba
e-mail: s1342034KU@s.chibakoudai.jp

T. Kohnosu
e-mail: tklab@it-chiba.ac.jp

T. Kobayashi
Nagasaki University, Nagasaki, Japan
e-mail: toru@cis.nagasaki-u.ac.jp

H. Sato
The University of Tokyo, Tokyo, Japan
e-mail: schuko@satolab.itc.u-tokyo.ac.jp

A. Kanai
Hosei University, Tokyo, Japan
e-mail: yoikana@hosei.ac.jp

R. Lee (ed.), *Software Engineering, Artificial Intelligence, Networking and Parallel/Distributed Computing*, Studies in Computational Intelligence 721,
DOI 10.1007/978-3-319-62048-0_8

Keywords Lifelog • Quantified self • Lifelog attribute service portfolio (LLADP) • Hybrid cloud computing • Data management • Data portfolio • Data privacy

1 Introduction

Lifelogs are now common in daily life, and many lifelogging technologies and applications have been designed for individuals to record/log their life events via personal computers and mobile devices. A new generation of applications and devices has been developed for individuals to compose digital autobiographies to recall events, review progress, and self-improve towards a better life. Typically, these technologies are incorporated into a smart wearable device for recording lifelog data such as personalized healthcare or wellness statistics. The 37° Bracelet, the Fitbit OneTM, SAGA, Google Glass, Google Fit, and tagCAM are examples of smart wearable devices/apps for wellness management; these lifelog/journal technologies were developed to constantly assist individuals in recording daily activities automatically as well as in capturing important or amazing moments in life. They also enable people to share details of their lifelog with friends, family, and followers in an open and closed form [1–3].

In addition, the potential exists for lifelogs to be used by employers to record/log the activities of employees. This potential is far into the future and should be addressed as an important issue for privacy concerns [2].

Furthermore, with regards to the challenges of lifelog technology, the massive amount of daily life data, the affordability of mass storage, and privacy concerns, we propose a set of attributes for lifelog data that can support and facilitate practical lifelogging for recording the entirety of daily life events. In addition, we propose a scheme for categorizing these lifelog attributes between private and public clouds (closed and open data), called a lifelog attribute data portfolio (LLADP), which takes into account individual privacy concerns.

This paper is organized as follows. Section 2 describes identifying lifelog events. A fundamental study of analysis of lifelog data portfolio is explained in Sect. 3. Specifically, the data structure of the life log was exhaustively extracted, and these data were analyzed in detail from the viewpoint of dynamic data attribute. Section 4 discusses consideration. Section 5 is the conclusion and future work.

2 Identifying Lifelog Events

A lifelog refers to the actual data captured using technology mostly based on wearable sensors. It gathers a user's life events and shares them with a selected audience. Generally, individuals digitally record their own information about health and fitness because they want to monitor and improve their behavior and wellbeing.

Technically and functionally, the lifelogging information derived from individual health information includes both physical activity data and emotional data, including heart rate, ambulatory blood pressure, respiratory rate, fatigue, sleep quality, public transport usage, regularity of lifestyle, etc. These data are also normally recorded with three sensing devices: GPS, accelerometers, and vital sign monitors [4–7].

The future challenge of lifelogging does not lie in health monitoring but in memory aiding. Lifelogging will be used for recording the entirety of life events with a series of photos and videos. The main role of a lifelog should be to record [8, 9].

1. Autobiographical memories: information, statistics, and data that help individuals in improving their lives.
2. Personal memories: daily life events in situations where individuals appear to be suffering from data fatigue.

Moreover, lower production costs for storage media offering more portability/storage capacity and the fast growth of wearable devices and mobile applications will create the biggest challenge. Individuals can record any memory or past experience by interacting with a digital device. The benefits of technology are to improve human life, relationships between people, and social interaction. On the other hand, certain recorded events may prove to be embarrassing.

Thus, the key question for these lifelog services is "what to log"? What kind of data should be kept secret, and what should be published? Cloud computing has now shifted to the practical use stage, and many cloud-related services increased sales in 2011. Moreover, many user companies are verifying the possibility and practicality of cloud computing in introducing information and communications technology (ICT). Cloud computing analysis is thus recognized as a key stage in systems configuration [10].

2.1 Reference Model of Lifelog Attribute Data Portfolio (LLADP)

In terms of quantified self-data modeling, Shimojo et al. [1]. Proposed the lifelog common data model (LLCDM)—the perspectives of what, why, who, when, where, and how—and also proposed the lifelog mashup API (LLAPI) for efficient integration of heterogeneous lifelog services for determining the different kinds of lifelog services. However, Kim and Giunchiglia stated in their research that a lifelog should contain "*a collection of digitized experiences from which a user may be able to retrieve much more detailed memories related with a particular event.*" [11]. Thus, what aspects of a particular event should be recorded have not been well established yet. As humans, we generally record our memories with five senses: sight, sound, touch, smell, and taste, whereas current lifelog technologies focus primarily on sight, with some consideration of sound [2].

For information capturing, Kim et al. [12]. Explained in their study that physical activities such as “standing,” “sitting,” “walking,” “running,” “lying,” “descending stairs,” “ascending stairs,” etc. are categorized by “time,” “location,” and a “pattern of consecutive behaviors.” Information within the “time” category is further categorized by “morning,” “afternoon,” or “evening.” The “location” category contains place names such as “school,” “office,” “home,” etc., whereas the “pattern of consecutive behaviors/object context” category contains information about the physical activities of individuals during daily life events.

Current wearable products and mobile applications have been released for the long term recording and collection of personal lifelogging physical activities (LPA). The LPA model consists of user movement activities recorded by GPS including location, distance, and speed [13]. Fitbit apps record and categorize daily activities as very active, moderately active, lightly active, and sedentary [6]. Other apps such as the 37° Bracelet track and record data such as steps and sleep: walking distance and caloric consumption. Moreover, GoLife, TiMeGo, and PAPAGO are apps that support route editing, road routing, and travel recording [14].

Based on these auto-recorded data, individuals can automatically log and record their physical, social, and daily activities, improve quality of life on the basis of daily activities, and plan life events such as sleeping, taking medication, allocating time for office tasks, or taking meals on the basis of their life-styles [15]. Besides those data, a lifelog service should support users in storing their entire daily life and in enabling them to manage and recall it easily. It should use photographs, physiological signals, time information, and the location of users. A lifelog service should help a user to recall and recognize behaviors the user was never aware of. It might also be used for decision making and personal service recommendations [1, 5, 13].

2.2 *Human Behavior Modeling*

Although lifelogging is a phenomenon whereby people digitally collect their life events using a lightweight neck-held camera and a smartphone with embedded GPS with varying levels of detail for various purposes, people generally use a wearable device and record their lifelog for the purposes of treating diseases and health problems or for improving their lives [2, 6, 8, 11, 16].

In addition, some use lifelogging as a memory aid, recording their life events such as those involving traveling and dining out and sharing them through text messages, photos, or videos via various applications.

In addition, in some cases, organizations may use lifelogging to record the activities of employees and consumer behavior to better understand their performance in a variety of tasks, replacing manual logging [2, 10, 12, 17].

In conclusion, a current lifelog service may have a data model (LLCDM) as recommended by Shimojo et al. [1]. From the perspectives of what, why, who, when, where, and how. However, patterns from the data stream—those that include aspects such as traditional diary entries—are mostly missing in current lifelog services. In addition, with the ease of capturing data and publishing them, new questions about the expectation of privacy are a hot topic [18].

Hence, we have redesigned the typical lifelog data service so that lifelog data streams have one distinct property, which is that they are a collection of events that occur in daily life. We describe the attributes for lifelog services on the basis of daily life events; that is, we create a data portfolio in a hybrid cloud with recommended levels of basic privacy (close/open) [19, 20].

3 Analysis of Lifelog Data Portfolio (LLADP)

In this section, to investigate an appropriate lifelog service data management method (i.e., a data portfolio) for a hybrid cloud configuration, we first find sources for lifelog data by reviewing previous literature as we did in Sect. 2. And we also try to use a work break down structure method (WBS), which is a typical method of estimation in project management, in defining the attribute service categories and their sub-coordinated attributes [1, 5, 13, 21].

3.1 Data Hierarchy of LLADP

In terms of quantified self data, we decide what we want to record, and we try to digitalize as much as what we can. For example, we record life activities as a visual sequence of digital images, personal biometrics, or communication activities [7]. Thus, in accordance with previous literature [12], we use a WBS for systematic organization of data and divide lifelog service categories into three main categories: “behavior log,” “vital log,” and “purchasing log,” as shown in Fig. 1.

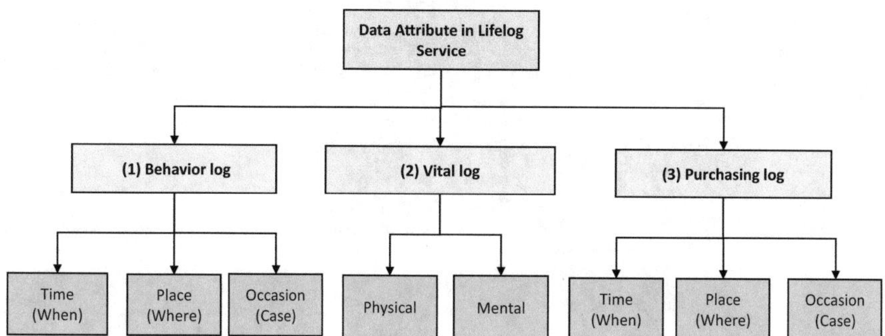

Fig. 1 First hierarchy of lifelog service data attributes

From Fig. 1, the "parent" branch of the hierarchy is the "behavior log," the "vital log," and the "purchasing log." Furthermore, the lower branches consist of "time," "location," and the "pattern of consecutive behaviors-occasion," etc.

3.2 Data Attributes of LLADP Guidelines and Compilation

Prior to Fig. 1, we identify behavior log attributes on the basis of daily life activities [11, 12, 15], as shown in Fig. 2. The first diagram describes the parent-child relationship of the "behavior" category.

In this study, we define "time," "place," and "occasion" at the second layer. In the third layer, a "child" branch of information recording, we divide the subcategories into two places: "home" and "office." In addition, we define the forth layer "time" for information recording into 2 periods: "morning" and "night." [12].

From Fig. 2, the "parent" branch of the hierarchy is "behavior," while the "child" branches consist of four layers:

- The first layer is "time," "place," and "occasion."
- The second layer is "home" and "office."
- The third layer is "morning" and "night."
- The fourth layer is the detailed attributes for "behavior."

For the "behavior log," we propose 27 items for recording in this service category.

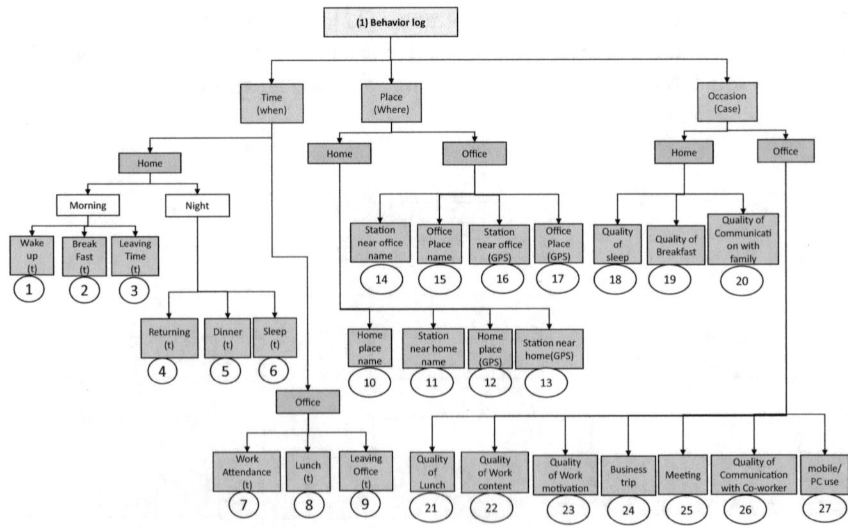

Fig. 2 Attributes of LLADP- "behavior log": 27 items

From Fig. 3, the "parent" branch of the hierarchy is the "vital log," while the "child" branches consist of two layers, "physical" and "mental" at the second layer, and the detailed attributes for "vital" at the third layer, which has 12 items for recording in this service category.

From Fig. 4, the "parent" branch of the hierarchy is the "purchasing log," while the "child" branches consist of two layers. The second layer consists of "time," "place," and "occasion." The third layer has six entities based on the second layer:

"Time" consists of two items, "purchase session" and "purchase time."

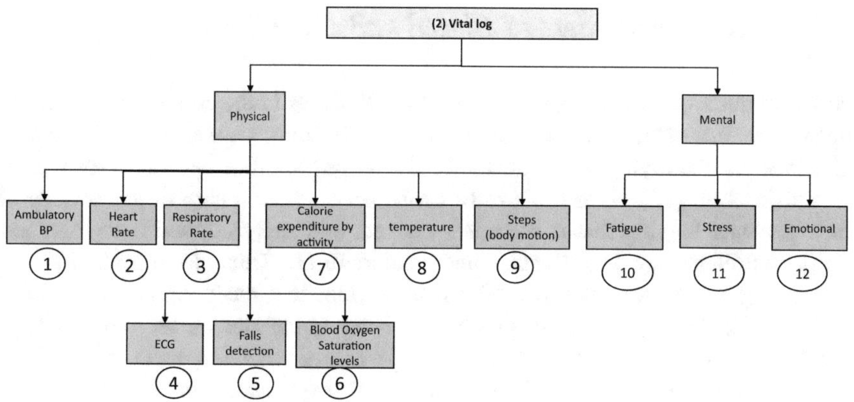

Fig. 3 Attributes of LLADP- "vital log": 12 items

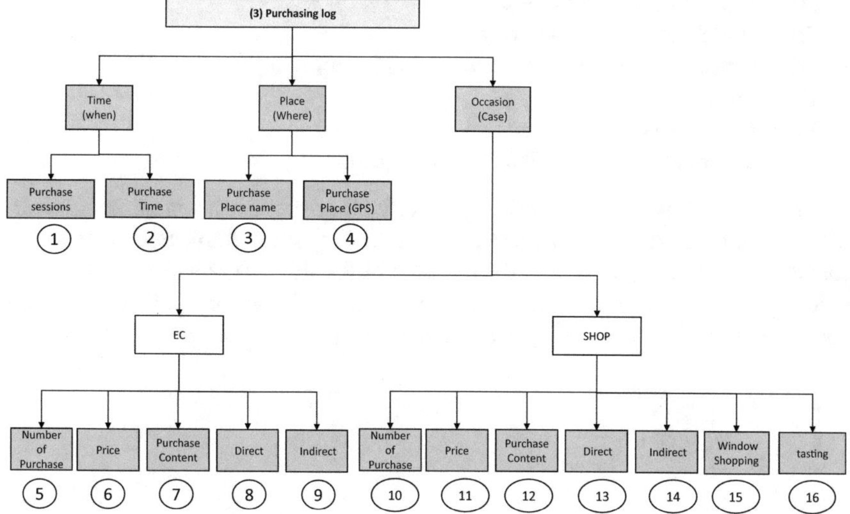

Fig. 4 Attributes of LLADP- "purchasing log": 16 items

"Place" also consists of two items, "purchase place name" and "purchase place (GPS information)."

"Occasion" consists of "E-commerce-EC" and "SHOP." The fourth layer is the detailed attributes for "occasion," which has 12 items for recording in this service category [1, 2, 17].

In this paper, from LLADP, we propose lifelog attributes for logging not only basic physical activities but also contexts based on a time series of individual life events [22].

3.3 LLADP in Context of Closed and Open Data

Next, we describe an application of the LLADP results in the context of closed and open data. For setting the criteria of privacy, we must establish rules to act as guidance for classifying which attributes should be closed or opened to the public.

Gurrin et al. [18]. Stated that most consumers are not aware of the privacy of their personal logging that is recorded, tracked, and analyzed, and even commercially distributed by broadcasters and related firms. Thus, for developing our lifelogging policy, we need to establish basic rules that meet expectations of privacy. In addition, Gurrin et al. proposed privacy guidelines on the basis of legal viewpoints and suggest that *"no one should record or photograph others for a lifelog without the consent of the person or their legal guardian."*

Furthermore, they suggested that the seven principles of lifelog privacy should meet the following conditions:

1. Proactive not reactive
2. Privacy as the default configuration
3. Privacy embedded into the design
4. Privacy as additional (not reduced) functionality
5. End-to-end data security
6. Visibility/transparency
7. Respect for the privacy of the individual user

For our work, it may be difficult to specify all privacy issues in all jurisdictions with one lifelog solution; however, we try to propose a framework and logical conditions for managing lifelog data to protect the life-logger and other actors by ensuring that his/her logging data complies with basic privacy policies. Our proposed decision flow based on privacy criteria is illustrated in Fig. 5 [23].

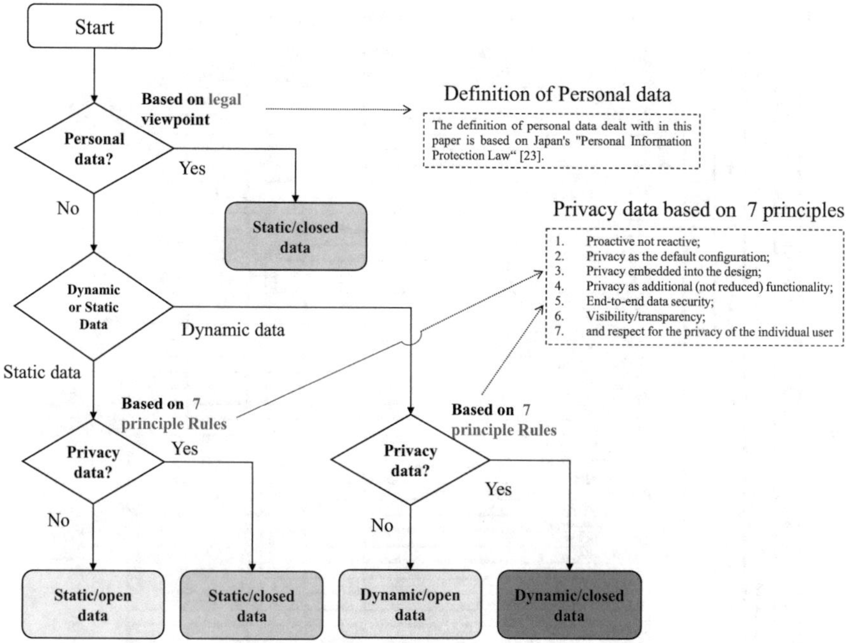

Fig. 5 Logical flowchart of privacy criteria to lifelog data portfolio

3.4 LLADP in Context of Static and Dynamic Data

The static data, once created, does not change; examples of static data for a lifelog are normally personal data such as an identification number, a social security number, and the life-logger's birth date and place of birth, whereas the dynamic lifelog data are created every day from wearable devices. The data are, for example, the tracked location, life-logger activity, and vital records of what we now call "quantified self" data. We apply this concept of static and dynamic data into our data portfolio management so that we can provide a scheme for arranging lifelog data between private and public clouds.

For both contexts of LLADP in these two dimensions, we propose a logical flow of privacy as shown in Fig. 5, and for lifelog data service management, by the data portfolio presented in Table 1.

Table 1 LLADP in context of closed and open, static and dynamic data

No	Level 1	Level 2	Level 3	Level 4	Level 5	Static Data: Open Data	Static Data: Closed data	Dynamic Data: Open Data	Dynamic Data: Closed data	Decision reason number in 7 principles
1	1. Behavior Log	1.1 Time	1.1.1 Home	1.1.1.1 Morning	1.1.1.1.1 Wake up time				O	
2					1.1.1.1.2 Breakfast time				O	
3					1.1.1.1.3 Leaving time				O	
4				1.1.1.2 Night	1.1.1.2.1 Returning time				O	
5					1.1.1.2.2 Dinner time				O	7
6					1.1.1.2.3 Sleep time				O	
7			1.1.2 Office		1.1.2.1.1 Work attendance time				O	
8					1.1.2.1.2 Lunch time				O	
9					1.1.2.1.3 Leaving office time				O	
10		1.2 Place	1.2.1 Home		1.2.1.1.1 Home place name		O			
11					1.2.1.1.2 Station near home name		O			
12					1.2.1.1.3 Home place GPS coordinates		O			
13					1.2.1.1.4 Station near home GPS coordinates		O			
14			1.2.2 Office		1.2.2.1.1 Station near office name		O			2, 5
15					1.2.2.1.2 Office place name		O			
16					1.2.2.1.3 Station near office GPS coordinates		O			
17					1.2.2.1.4 Office place GPS coordinates		O			
18		1.3 Occasion	1.3.1 Home		1.3.1.1.1 Quality of sleep				O	
19					1.3.1.1.2 Quality of breakfast				O	7
20					1.3.1.1.3 Quality of communication with family				O	
21			1.3.2 Office		1.3.2.1.1 Quality of lunch				O	
22					1.3.2.1.2 Quality of work content		O			
23					1.3.2.1.3 Quality of work motivation				O	7
24					1.3.2.1.4 Business trips		O			
25					1.3.2.1.5 Meetings		O			
26					1.3.2.1.6 Quality of communication with co-workers				O	7
27					1.3.2.1.7 Mobile/PC use		O			
28	2. Vital Log	2.1 Physical			2.1.1.1.1 Ambulatory BP		O			2, 5
29					2.1.1.1.2 Heart rate		O			
30					2.1.1.1.3 Respiratory rate		O			
31					2.1.1.1.4 ECG		O			
32					2.1.1.1.5 Fall detection		O			
33					2.1.1.1.6 Blood oxygen saturation levels		O			
34					2.1.1.1.7 Calorie expenditure by activity		O			
35					2.1.1.1.8 Temperature		O			
36					2.1.1.1.9 Steps (body motion)				O	
37		2.2 Mental			2.2.1.1.1Fatigue				O	
38					2.2.1.1.2 Stress				O	
39					2.2.1.1.3 Emotional state				O	
40	3. Purchasing log	3.1 Time			3.1.1.1.1 Purchase sessions				O	
41					3.1.1.1.2 Purchase time				O	
42		3.2 Place			3.1.1.2.1 Purchase place name				O	
43					3.1.1.2.2 Purchase place GPS coordinates				O	
44		3.3 Occasion		3.3.1.1 EC e-Commerce	3.3.1.1.1Number of purchases				O	
45					3.3.1.1.2 Price				O	
46					3.3.1.1.3 Purchase content				O	7
47					3.3.1.1.4 Direct				O	
48					3.3.1.1.5 Indirect				O	
49				3.3.1.2 Shop	3.3.1.2.1 Number of purchases				O	
50					3.3.1.2.2 Price				O	
51					3.3.1.2.3 Purchase content				O	
52					3.3.1.2.4 Direct				O	
53					3.3.1.2.5 Indirect				O	
54					3.3.1.2.6 Window shopping				O	
55					3.3.1.2.7 Tasting				O	
					Subtotal of an corresponding item	0	20	0	35	

4 Discussion

This study proposed a lifelog attribute service portfolio that classifies lifelog data from continuous data streams of user life event activities and developed a portfolio called the "LifeLog Attribute Data Portfolio-LLADP." We then classified the attributes into static-closed, static-open, and dynamic-closed/open categories for hybrid cloud data management and for the protection of individual privacy.

Figure 6 shows a summary of the results. The results of the data portfolio show that privacy concerns are an important factor when classifying life activities. Thus, it was generally proved that lifelog data are closed data. However, it also became

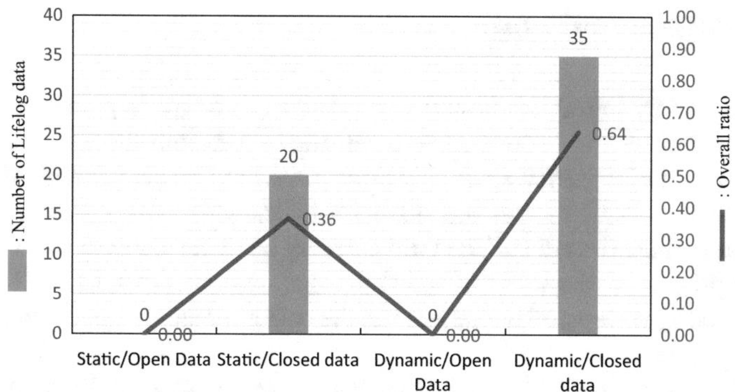

Fig. 6 Result summary of LLADP

clear from the viewpoint of data attributes that about 60% are dynamic data attributes.

In these dynamic data attributes, cases where closed data will gradually change to open data are assumed. For example, it is a case that "closed data" changes to "open data" due to individual attribute change (employee ⇒ retirement etc.).

Thus, it shows that a data portfolio with consideration for such data attributes is important.

This portfolio will be used to establish basic guidelines for hybrid cloud data management from a practical viewpoint. The closed data should be kept in a private cloud, and the open data should be kept in a public cloud, as shown in Fig. 7 [24].

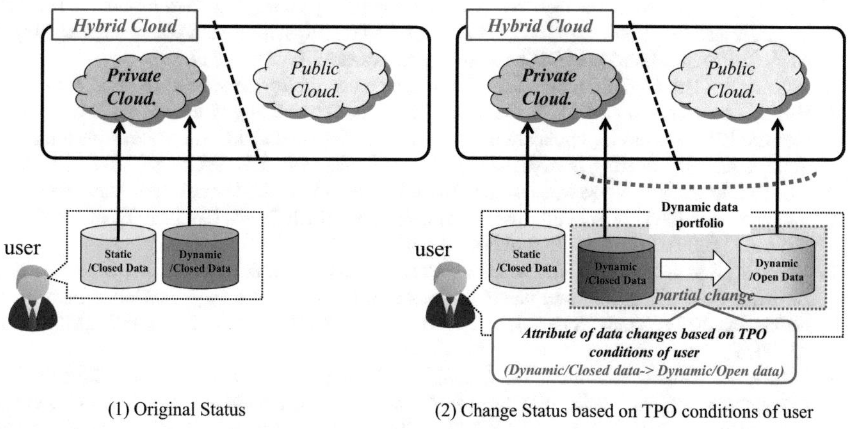

Fig. 7 Reference model of hybrid cloud computing [24]

5 Conclusion and Future Work

This paper mainly introduces how data attributes can be made public or private in accordance with privacy concerns. We categorize data into three types. The first category is for attributes that are highly private and must be kept closed. The second category is for attributes that are not so private, where users can always give permission to share these data. The third category is for attributes that can be judged public or private on the basis of individual permissions. This LLADP will help users to enact a privacy policy that enables the data attributes to appear in the lifelog of any life-logger.

Our future work is to examine the dynamic change of these attributes on the basis of average information content (entropy). Future experiments will use open real data, and further evaluations will be performed. In addition, the final results will be used to classify a lifelog privacy framework for each individual in regards to the data organization challenges and the data presentation challenges of each attribute.

Acknowledgements This work was supported by the Japan Society for the Promotion of Science (JSPS, KAKENHI Grant Number 15H02783).

References

1. Shimojo, A., Kamada, S., Matsumoto, S., Nakamura, M.: On integrating heterogeneous lifelog services. In: Proceedings of the 12th International Conference on Information Integration and Web-based Applications & Services, Paris, pp. 263–272, Nov 2010. doi:10.1145/1967486.1967529
2. Gurrin, C., Smeaton, A.F., Doherty, A.R.: Lifelogging: personal big data. In: Found Trends® in Information Retrieval **8**(1), 1–125 (2014). https://pdfs.semanticscholar.org/0246/edc6c018faa245e1ec88ba0fe916be2c3081.pdf. Accessed 06 Mar 2017
3. Ireland, K.: The Digital Divide. http://www.getsaga.com/blog/. Accessed 09 Mar 2017
4. Ho, Y. sato-Shimokawara, E., Yamaguchi, T.: (2012) Data Mining of life log for developing a user model-based service application. In: 2012 IEEE International Conference on Automation Science and Engineering (CASE), Seoul, pp. 757–760. doi:10.1109/CoASE.2012.6386491
5. Abe, M., Morinishi, Y., Maeda, A., Aoki, M., Inagaki, H.: A life log collector integrated with a remote-controller for enabling user centric services. IEEE Trans. Consum. Electron. **55**(1), 295–302 (2009). doi:10.1109/TCE.2009.4814448
6. Yang, P., Stankevicius, D., Marozas, V., Deng, Z., Liu, E., Lukosevicius, A., Dong, F., Xu, L., Min, G.: Lifelogging data validation model for internet of things enabled personalized healthcare. IEEE. Trans. Syst. Man Cybern.: Syst. **PP**(99), 1–15. doi:10.1109/TSMC.2016.2586075
7. Duane, A., Gupta, R., Zhou, L., Gurrin, C.: Visual insights from personal lifelogs. In: Proceedings of the 12th NTCIR Conference on Evaluation of Information Access Technologies, Tokyo, pp. 386–389 (2016). http://research.nii.ac.jp/ntcir/workshop/OnlineProceedings12/pdf/ntcir/LIFELOG/08-NTCIR12-LIFELOG-DuaneA.pdf. Accessed 10 Mar 2017
8. Rawassizadeh, R., Tomitsch, M., Wac, K., Tjoa, A.M.: UbiqLog: a generic mobile phone-based life-log framework. Pers. Ubiquit. Comput. **17**(4), 621–637 (2013). doi:10.1007/s00779-012-0511-8

9. Machajdik, J., Hanbury, A., Garz, A., Sablatnig, R.: Affective computing for wearable diary and lifelogging systems: an overview. In: Machine Vision-Research for High Quality Processes and Products-35th Workshop of the Austrian Association for Pattern Recognition. Austrian Computer Society (2011). http://allan.hanbury.eu/lib/exe/fetch.php?media=machajdik_aapr_2011.pdf. Accessed 10 Mar 2017
10. Boldt, L.C., et al.: Forecasting Nike's sales using Facebook data. In: 2016 IEEE International Conference on Big Data (Big Data), Washington, DC, pp. 2447–2456 (2016). doi:10.1109/BigData.2016.7840881
11. Kim, P.H., Giunchiglia, F.: Life logging practice for human behavior modeling. In: 2012 IEEE International Conference on Systems, Man, and Cybernetics (SMC), Seoul, pp. 2873–2878 (2012). doi:10.1109/ICSMC.2012.6378185
12. Kim, M., Lee, D.W., Kim, K., Kim, J.H., Cho, W.D.: Predicting personal information behaviors with lifelog data. In: 2012, 9th International Conference & Expo on Emerging Technologies for a Smarter World (CEWIT), Incheon, pp. 1–3 (2012). doi:10.1109/CEWIT.2012.6606983
13. Ryoo, D.W., Bae, C.: Design of the wearable gadgets for life-log services based on UTC. IEEE Trans. Consum. Electron. **53**(4), 1477–1482 (2007). doi:10.1109/TCE.2007.4429240
14. Huang, F.M., Huang, Y.H., Szu, C., Su, A.Y.S., Chen, M.C., Sun, Y.S.: A study of a life logging smartphone app and its power consumption observation in location-based service scenario. In: 2015 IEEE International Conference on Mobile Services, New York, NY, pp. 225–232 (2015). doi:10.1109/MobServ.2015.40
15. Jalal, A., Kamal, S.: Real-time life logging via a depth silhouette-based human activity recognition system for smart home services. In: 2014 11th IEEE International Conference on Advanced Video and Signal Based Surveillance (AVSS), Seoul, pp. 74–80 (2014). doi:10.1109/AVSS.2014.6918647
16. Ushiama, T., Watanabe, T.: A life-log search model based on Bayesian network. In: IEEE Sixth International Symposium on Multimedia Software Engineering, pp. 337–343. doi:10.1109/MMSE.2004.11 (2004), Huang, C.L., Huang, Y.H., Chen, J.J.: Life events segmentation based on lifelog recorded by wearable device. In: 2015 International Conference on Intelligent Information Hiding and Multimedia Signal Processing (IIH-MSP), Adelaide, SA, pp. 129–132 (2015). doi:10.1109/IIH-MSP.2015.19
17. Lin, Y.-T.J, Lin, M.-Y.T, Li, K.-C.: Consumer involvement model of fan page: mining from Facebook data of a real celebrity fashion brand. In: 2015 12th International Conference on Service Systems and Service Management (ICSSSM), Guangzhou, pp. 1–6 (2015). doi:10.1109/ICSSSM.2015.7170187
18. Gurrin, C., Albatal, R., Joho, H., Ishii, K.: A privacy by design approach to lifelogging. In: Digital Enlightenment Yearbook, pp. 49–73 (2014). http://doras.dcu.ie/20505/1/Gurrin.pdf. Accessed 09 Mar 2017
19. Beck, M., Hao, W., Campan, A.: Accelerating the mobile cloud: using amazon mobile analytics and k-means clustering. In: 2017 IEEE 7th Annual Computing and Communication Workshop and Conference (CCWC), Las Vegas, NV, USA, pp. 1–7 (2017). doi:10.1109/CCWC.2017.7868372
20. Huang, C.L., Huang, Y.H., Chen, J.J.: Life events segmentation based on lifelog recorded by wearable device. In: 2015 International Conference on Intelligent Information Hiding and Multimedia Signal Processing (IIH-MSP), Adelaide, SA, pp. 129–132 (2015). doi:10.1109/IIH-MSP.2015.19
21. Project Management Institute: A Guide to the Project Management Body of Knowledge. http://www.cs.bilkent.edu.tr/~cagatay/cs413/PMBOK.pdf. Accessed 23 Apr 2017
22. Giunchiglia, F., Kim, P.H.: Lifelog data model and management: study on research challenges. Int. J. Comput. Inf. Syst. Industr. Manage. Appl. **5**, 115–125 (2012). ISSN 2150-7988

23. Personal Information Protection Commission: About personal information protection law. https://www.ppc.go.jp/personal/general/ (in Japanese). Accessed 23 Apr 2017
24. Tanimoto, S., Sakurada, Y., Seki, Y., Iwashita, M., Matsui, S., Sato, H., Kanai, A.: A study of data management in hybrid cloud configuration. In: 14th IEEE/ACIS, SNPD2013, pp. 381–386

Estimation of Emotional Scene from Lifelog Videos in Consideration of Intensity of Various Facial Expressions

Shota Sakaue, Hiroki Nomiya and Teruhisa Hochin

Abstract Previous research has proposed facial expression intensity to estimate the intensity of facial expression for the purpose of retrieving impressive scenes from lifelog videos. However, only the intensity of the smile and that of the surprise are estimated. We propose the improved method of obtaining facial expression intensity, and estimate emotional scenes based on the basic six facial expressions from lifelog videos. We evaluate the proposed method by using a commonly-used facial expression data set and lifelog videos. Then, we compare the result of the experiment with the previous research. The proposed method is more effective than the previous one for the detection of the impressive scenes with smile from lifelog videos. It is also confirmed that the proposed method can precisely estimate facial expression intensity for the other facial expressions.

Keywords Lifelog · Video retrieval · Facial expression recognition · Emotion

1 Introduction

In recent years, because of improvement of the performance of multimedia recording devices and storage device and reduction of the price, anyone can easily create a large quantity of multimedia data. For such a reason, *lifelog* has been proposed for getting and storing a variety of events in daily life as various types of data [1–3].

S. Sakaue
Graduate School of Information Science, Kyoto Institute of Technology,
Goshokaido-cho, Matsugasaki, Sakyo-ku, Kyoto 606-8585, Japan
e-mail: m5622013@edu.kit.ac.jp

H. Nomiya (✉) · T. Hochin
Information and Human Sciences, Kyoto Institute of Technology, Goshokaido-cho,
Matsugasaki, Sakyo-ku, Kyoto 606-8585, Japan
e-mail: nomiya@kit.ac.jp

T. Hochin
e-mail: hochin@kit.ac.jp

R. Lee (ed.), *Software Engineering, Artificial Intelligence, Networking and Parallel/Distributed Computing*, Studies in Computational Intelligence 721,
DOI 10.1007/978-3-319-62048-0_9

Particularly, video data can be created easily and contain a variety of useful information. Therefore, we focus on lifelog videos in this study.

Lifelog videos have been created easily, and a large amount of video data can be stored in the video databases. This makes the retrieval of lifelog videos quite difficult. Consequently, a considerable amount of valuable lifelog data is not utilized. Hence, anyone wants to efficiently and accurately search useful video scenes in the lifelog videos.

For the better utilization of the lifelog videos, an effective retrieval method has been proposed [4–6]. This method detects impressive scenes from a lifelog video by using the facial expressions of a person in the video. It can retrieve emotional scenes that a person expresses a kind of facial expression. However, the intensity of the emotion cannot be estimated. For example, it can detect "smile" but it cannot distinguish "giggle" from "laughter."

Morikuni et al. thought that a number of emotional scenes expressing various levels of facial expression intensity are included in lifelog videos, and they proposed a measure of "facial expression intensity" for emotional scene retreival in lifelog videos, which is the criterion to measure the intensity of facial expression [7]. The facial expression intensity is calculated using several salient points on a face called *facial feature points*.

However, this method did not evaluate facial feature values for calculating facial expression intensity, and there was still a problem in estimation accuracy. Furthermore, it focused only on "happiness" and "surprise," and did not provide any estimation methods for facial expressions such as "anger" and "sadness," which are relatively difficult to estimate.

In this paper, we propose new facial feature values to estimate the intensity of basic six facial expressions (anger, disgust, fear, happiness, sadness, and surprise), and then we evaluate the facial feature values. We propose a new facial expression intensity with effective facial feature values. Also, we prepare new lifelog videos, and estimate emotional scenes in the lifelog videos with the proposed facial expression intensity. We show that the proposed method is superior to the previous one.

The remainder of this paper is organized as follows. Section 2 presents the previous method. Section 3 describes new facial feature values and the calculation of the facial expression intensity using the proposed facial feature values. Section 4 shows the experiment to evaluate the usefulness of the facial expression intensity and the result of experiment. Section 5 gives consideration to the result of expriment. Finally, Sect. 6 concludes this paper.

2 Previous Research

Morikuni et al. proposed facial expression intensity for the purpose of expressing the degree of facial expressions [7]. In this section, we explain the facial expression intensity. And then, we show the previous method which calculates the facial expression intensity with several facial feature values.

2.1 Facial Expression Intensity

In most of conventional researches, it is possible to recognize the type of facial expressions, but could not estimate the degree of facial expressions. Therefore, Morikuni et al. focused on the fact that the larger the degree of facial expressions is, the larger the facial expressions change. They proposed facial expression intensity, which is a criterion to measure the intensity of the emotion on the basis of the movement of salient facial feature points [7]. The movement of the facial feature points is represented by some facial feature values defined as the positional relationships of the facial feature points.

2.1.1 Facial Feature Points

The facial feature points are extracted by using Luxand FaceSDK4.0 [8]. The facial expression intensity is calculated based on the positional relationships of the extracted fifty four facial feature points. We show the number of facial feature points at each part of a face as follows. Figure 1 shows the details of facial feature points.

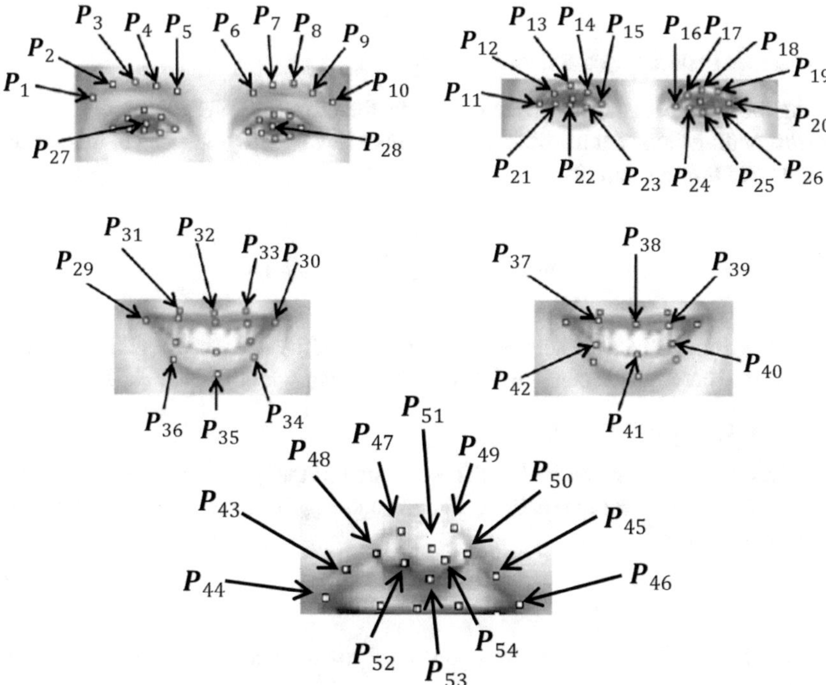

Fig. 1 Details of facial feature points

- Left and right eyebrows: 10 points ($P_1, \ldots, P_{10}$)
- Left and right eyes: 18 points ($P_{11}, \ldots, P_{28}$)
- Mouth: 14 points ($P_{29}, \ldots, P_{42}$)
- Nasolabial folds: 4 points ($P_{43}, \ldots, P_{46}$)
- Nose: 8 points ($P_{47}, \ldots, P_{54}$)

2.1.2 Facial Feature Value

Morikuni et al. defined the following 11 types of facial feature values to estimate the facial expression intensity [7]. These facial feature values are considered to be associated with the change in facial expressions.

- Slopes of right and left eyebrows: f_1

This feature value is based on the slopes of the right and left eyebrows (denoted by a_r and a_l, respectively) calculated by using least squares. The slopes of the left and right eyebrows are obtained from the facial feature points $P_6, P_7, P_8, P_9, P_{10}$, ($P_1, P_2, P_3, P_4, P_5$, respectively). This feature value is defined as Eq. (1).

$$f_1 = \frac{a_l - a_r}{2} \tag{1}$$

- Distance between eyes and eyebrows: f_2

Using the average distance between ten facial feature points on the eyebrows and the corresponding facial feature points on the upper side of eyes, the value of this feature is obtained through Eq. (2).

$$f_2 = \frac{1}{10 \cdot L} \cdot \sum_{i=1}^{10} ||P_i - P_{i+10}|| \tag{2}$$

Here, L is a normalization factor for the difference of the size of a face. It is defined as the distance between the center points of left and right eyes, that is, $L = ||P_{27} - P_{28}||$.

- Area between eyebrows: f_3

This feature value is given by Eq. (3) as the area of the quadrangle formed by connecting four facial feature points P_5, P_6, P_{16}, and P_{15} located at the inner corners of eyebrows and eyes.

$$f_3 = \frac{S(P_5, P_6, P_{16}, P_{15})}{L^2} \tag{3}$$

Here, $S(P_1, \ldots, P_n)$ is the area of a polygon formed by connecting n facial feature points $P_1, \ldots, P_n$, and L^2 is a normalization factor for the difference of the size of a face.

- Area of eyes: f_4

This feature value is the averaged areas of two octagons formed by the facial feature points on the circumference of the left and right eyes, defined by Eq. (4).

$$f_4 = \frac{1}{2 \cdot L^2} \begin{Bmatrix} S(P_{11}, P_{12}, P_{13}, P_{14}, P_{15}, P_{23}, P_{22}, P_{21}) \\ +S(P_{16}, P_{17}, P_{18}, P_{19}, P_{20}, P_{26}, P_{25}, P_{24}) \end{Bmatrix} \tag{4}$$

- Vertical-to-horizontal ratio of eyes: f_5

Based on the ratio of the distance between the top and bottom points on the left and right eyes (denoted by l_{vl} and l_{vr}, respectively) to the distance between the left and right points on the left and right eyes (denoted by l_{hl} and l_{hr}, respectively), this feature value is defined by Eq. (5).

$$f_5 = \frac{1}{2}(\tan^{-1}\frac{l_{vl}}{l_{hl}} + \tan^{-1}\frac{l_{vr}}{l_{hr}}) \tag{5}$$

Here, l_{vl}, l_{vr}, l_{hl} and l_{hr} are defined by Eq. (6).

$$\begin{cases} l_{vl} = ||P_{25} - P_{18}|| \; l_{hl} = ||P_{15} - P_{11}|| \\ l_{vr} = ||P_{22} - P_{13}|| \; l_{hr} = ||P_{20} - P_{16}|| \end{cases} \tag{6}$$

- Area of inner circumference of a mouth: f_6

This feature value is defined by Eq. (7) as the area of the octagon formed by connecting eight facial feature points located on the inner circumference of a mouth.

$$f_6 = \frac{S(P_{29}, P_{37}, P_{38}, P_{39}, P_{30}, P_{40}, P_{41}, P_{42})}{L^2} \tag{7}$$

- Area of outer circumference of a mouth: f_7

Similar to the sixth feature value, this feature value is defined by Eq. (8) as the area of the octagon formed by connecting eight facial feature points located on the outer circumference of a mouth.

$$f_7 = \frac{S(P_{29}, P_{31}, P_{32}, P_{33}, P_{30}, P_{34}, P_{35}, P_{36})}{L^2} \tag{8}$$

The sixth feature value are influenced by the thickness of the lips which can vary depending on the type and the intensity of the facial expression. On the other hand, this feature value is hardly affected by the thickness of the lips.

- Vertical-to-horizontal ratio of the inner circumference of a mouth: f_8

Based on the ratio of the distance m_{iv} between the top and bottom points on the inner circumference of a mouth to the distance m_{ih} between the left and right points on the inner circumference of a mouth, this feature value is defined by Eq. (9).

$$f_8 = \tan^{-1}\frac{m_{iv}}{m_{ih}} \tag{9}$$

Here, m_{iv} and m_{ih} are defined by Eq. (10).

$$\begin{cases} m_{iv} = ||P_{41} - P_{38}|| \\ m_{ih} = ||P_{30} - P_{29}|| \end{cases} \tag{10}$$

- Vertical-to-horizontal ratio of the outer circumference of a mouth: f_9

Similar to the eighth feature value, this feature value is defined by Eq. (11) based on the ratio of the distance m_{ov} between the top and bottom points on the outer circumference of a mouth to the distance m_{oh} between the left and right points on the outer circumference of a mouth. This feature value is also insensitive to the thickness of the lips.

$$f_9 = \tan^{-1}\frac{m_{ov}}{m_{oh}} \tag{11}$$

Here, m_{ov} and m_{oh} are defined by Eq. (12).

$$\begin{cases} m_{ov} = ||P_{35} - P_{32}|| \\ m_{oh} = ||P_{30} - P_{29}|| \end{cases} \tag{12}$$

- Vertical position of the corner of a mouth: f_{10}

This feature value represents how high the position of the corner of a mouth is. It is defined by Eq. (13).

$$f_{10} = \frac{(y(P_{29}) + y(P_{30})) - (y(P_{32}) + y(P_{35}))}{||y(P_{32}) + y(P_{35})||} \tag{13}$$

Here, $y(P)$ is the y-coordinate of the facial feature point P. If the mean value of the y-coordinate of the facial feature points on the corner of a mouth is larger than that of the facial feature points on the top and bottom of a mouth, f_{10} becomes positive. Thus, a larger value of f_{10} represents a higher vertical position of the corner of a mouth.

- Angles of corners of a mouth: f_{11}

This feature value is the mean value of the angles of the left and right corners of a mouth. The angle of the left (right, respectively) corner is formed by connecting the three facial feature points located on the left (right) corner of a mouth. It is given by Eq. (14).

$$f_{11} = \frac{A(P_{29}, P_{31}, P_{36}) + A(P_{30}, P_{33}, P_{34})}{2} \tag{14}$$

Here, $A(p, q, r)$ is the function to calculate the angle formed by three facial feature points $p, qandr$. A is defined by Eq. (15).

$$A(p,q,r) = \cos^{-1} \frac{(p-q)\cdot(p-r)}{||p-q||\cdot||p-r||} \tag{15}$$

For each frame in a lifelog video, the above facial feature values are calculated and the feature vector $(f_{m1}, \ldots, f_{m11})$ is obtained. Here, the j-th facial feature value obtained from the m-th frame from the beginning of the lifelog video is denoted by f_{mj}.

2.1.3 Expression Intensity

The facial expression intensity is defined for a single frame taking into consideration the tendency that the facial feature values are proportional to the intensity of emotion. In order to accurately estimate the intensity of emotion, the baseline of facial feature values are first determined using a training data set. The training data set consists of twelve emotional frames and twelve nonemotional frames. An emotional frame is the frame that a person in the frame image expresses a certain emotion. On the other hand, a nonemotional frame is the frame that the facial expression of a person in the frame is neutral.

The training set is manually and subjectively prepared prior to the calculation of the facial expression intensity. The training data set is represented as $T = \{g_1, \ldots, g_{11}, G_1, \ldots, G_{11}\}$. Here, g_j and $G_j(j = 1, \ldots, 11)$ are the mean value of facial feature values of the emotional and nonemotional frames, respectively.

Because of the personal difference of the facial expressions, it is quite difficult to estimate the facial expression intensity directly from the facial feature values. Hence, the baseline of facial feature values are calculated to diminish the personal difference of the facial feature values.

The baseline facial feature value is determined for each facial feature. The baseline of the j-th facial feature $S_j(j = 1, \ldots, 11)$ is defined by Eq. (16).

$$S_j = \frac{g_j + G_j}{2} \tag{16}$$

The facial expression intensity is calculated on the basis of the difference between the feature values obtained from a frame and the baseline feature values. the facial expression intensity of the m-th frame in a lifelog video E_m is defined by Eq. (17).

$$E_m = \sum_{j=1}^{11} (f_{mj} - S_j) \tag{17}$$

Therefore, the higher value of the expression intensity represents the stronger emotion.

3 Proposed Method

In the previous research, the evaluation of the facial feature values are not carried out. There is also a problem that the retrieval accuracy is poor. In this section, in order to improve the retrieval accuracy, we propose new facial feature values. It was evaluated with the existing facial feature values.

3.1 New Facial Feature Values

We add five new facial features to the set of facial features defined in the previous study [7]. One of the authors analyzed the positional relationships of the facial feature points in a number of facial images and subjectively determined the following new facial features which were likely to enhance the estimation ability of the facial expression intensity.

- Slopes of right and left nasolabial folds: f_{12}

Similar to the first feature value, this feature value is defined by Eq. (18) based on the slopes of the right and left nasolabial folds (denoted by b_r and b_l, respectively) calculated by using least squares. The slopes of the left and right nasolabial folds are obtained from the facial feature points $P_{45}, P_{46}, P_{49}, P_{50}$ and $P_{43}, P_{44}, P_{47}, P_{48}$, respectively.

$$f_{12} = \frac{b_l - b_r}{2} \tag{18}$$

- Area of a triangle that connects both ends of a mouth and a nose apex: f_{13}

This feature value is defined by Eq. (19) as the area of the triangle formed by connecting three facial feature points P_{29}, P_{30}, and P_{51} located on both ends of a mouth and a nose apex.

$$f_{13} = \frac{S(P_{29}, P_{30}, P_{51})}{L^2} \tag{19}$$

- Distance between a nose and a mouth: f_{14}

Similar to the second feature value, this feature value is defined by Eq. (20) using the average distance between three facial feature points on the bottom of a nose and the corresponding facial feature points on the upper side of a mouth.

$$f_{14} = \frac{1}{3 \cdot L} \cdot \sum_{i=31}^{33} ||P_i - P_{i+21}|| \tag{20}$$

- Area of a quadrangle that connects the bottom of nasolabial fold points and both ends of a mouth: f_{15}

This feature value is defined by Eq. (21) as the area of the quadrangle formed by connecting four facial feature points located on the bottom of nasolabial fold points and both ends of a mouth.

$$f_{15} = \frac{S(P_{29}, P_{30}, P_{46}, P_{43})}{L^2} \tag{21}$$

- Area of quadrangles that connects at both ends of eyes and both ends of eyebrows: f_{16}

Similar to the fifteenth feature value, this feature value is defined by Eq. (22) as the mean of the area of the two quadrangles formed by connecting four facial feature points located on both ends of left and right eyes and both ends of left and right eyebrows.

$$f_{16} = \frac{1}{2 \cdot L^2} \cdot \left\{ \begin{array}{l} S(P_1, P_{11}, P_{15}, P_5) \\ +S(P_6, P_{16}, P_{20}, P_{10}) \end{array} \right\} \tag{22}$$

3.2 Evaluation of Facial Feature Values

In the proposed method, the feature values are evaluated, and the feature values for calculating facial expression intensity are individually determined for each facial expression. First, in each facial expression, a change amount in the feature value and a change amount in the facial expression are visually compared in a lifelog video. In this paper, one of the authors compares them subjectively. As a result of the comparison, if it is considered that there is a positive correlation between them, the facial feature value is added when obtaining the facial expression intensity. Likewise, if there is a negative correlation, it is subtracted. If there is no correlation, the feature value is not used because it is considered to be ineffective.

3.3 Facial Feature Values in Each Facial Expression

Here, we describe the feature values used to calculate the facial expression intensity. The facial feature values determined to be usable for the estimation of the intensity of each facial expression are shown in Table 1. The facial feature values in parentheses represent the feature values that have negative correlation. For details of the evaluation, please refer to [9].

3.4 Proposed Expression Intensity

Here, we propose the method to calculate the facial expression intensity. First, we standardize the evaluated feature values. We define the *j*-th standardized facial fea-

Table 1 The facial feature values used for estimation of expression intensity

Facial expression	Facial feature values
Anger	f_1, (f_2,f_3,f_4,f_5)
Disgust	f_1,f_{10},f_{11}, (f_2,f_3,f_4,f_5,f_{16})
Fear	f_2,f_3,f_4,f_5, (f_1,f_{13},f_{14})
Happiness	f_{10},f_{12}
Sadness	f_2,f_3,f_9,f_{13},f_{15},f_{16}, (f_4,f_5,f_{10})
Surprise	f_2,f_{16}, (f_1,f_3)

ture value of the m-th frame in a lifelog video as F_{mj} ($-F_{mj}$, respectively) when there is a positive (negative) correlation. Second, in the same way as in the previous research, we select training data and calculate the mean value of facial feature values of emotional and nonemotional frames as $g'_1, \ldots, g'_{16}$ and $G'_1, \ldots, G'_{16}$ using F_{mj} and $-F_{mj}$. The baseline of the j-th facial feature $S'_j (j = 1, \ldots, 16)$ is defined by Eq. (23).

$$S'_j = \frac{g'_j + G'_j}{2} \quad (j = 1, \ldots, 16) \tag{23}$$

Finally, the expression intensity of the m-th frame in a lifelog video E'_m is defined by Eq. (24).

$$E'_m = \sum_{j=1}^{16} \begin{cases} (F_{mj} - S'_j) & (\textit{positive correlation}) \\ (-F_{mj} - S'_j) & (\textit{negative correlation}) \\ 0 & (\textit{no correlation}) \end{cases} \tag{24}$$

4 Experiment

We estimate facial expression intensity and emotional scenes in the acquired lifelog videos and a facial expression dataset.

4.1 Datasets

4.1.1 Lifelog Videos

We prepared lifelog videos including the scenes that the subjects play games, and used them as a lifelog video dataset. The subjects are two male graduate students (subjectA, B) including one of the authors, and four female undergraduate students (subjectC, …, subjectF), eighteen to twenty six years old. We estimate facial expression intensity using the lifelog videos. In the previous method, in order to evaluate the estimation accuracy of facial expression intensity, Morikuni et al. classified the

Table 2 The judgement criteria for each intensity level

Intensity level	Criteria
Happiness1	A mouth does not open much, both ends of lip are raised
Happiness2	The corners of the eyes drop, opening a mouth and raise a laughter
Happiness3	The corners of the eyes drop, opening the mouth widely, laughing with loud voice and gestures
Surprise1	A surprise as happens when you notice something or when you admire it
Surprise2	Surprise to something, a surprise like opening mouth
Surprise3	A surprise like bending backward and opening the mouth wide

happiness into three levels according to the intensity of the happiness [7]. The reason for this classification is that the correct facial expression intensity must manually be determined to evaluate the estimation accuracy of the facial expression intensity, but it is very difficult to manually determine the correct facial expression intensity in a real number. In the proposed method, one of the authors classified the intensity of "surprise" into three levels as shown in Table 2. Note that the criteria for "happiness1" to "happiness3" are the same as defined in [7]. The classification of the intensity of facial expressions shown in Table 2 will be referred to as the "intensity level."

4.1.2 MMI Datasets

In this experiment, we use the dataset in the MMI facial expression database [10, 11]. This dataset contains six basic facial expressions. Among them, we estimate the facial expression intensity for "anger," "disgust," and "sadness" because these facial expressions are not observed in the lifelog dataset. This dataset is composed of videos of a facial expression of each subject. These videos start with a neutral face. After facial expression changes gradually, it appears completely. Then it ends with a neutral face again. In this experiment, we calculate the facial expression intensity of anger (5 subjects), disgust (8 subjects), and sadness (4 subjects). One of the authors selected these samples in which the facial feature points were accurately extracted.

4.1.3 Training and Test Sets

The emotional frames in the training sets of lifelog videos and MMI datasets were subjectively determined by one of the authors. This is the same method used in the previous study [7]. The test sets were subjectively determined so that the frames in a test set were not included in the video from which the frames in a training set were selected. The frames in the training and test sets were selected from the emotional frames with accurate facial feature points. This is also the same way used in the previous study.

4.2 Experimental Result

4.2.1 Lifelog Videos

Table 3 shows the mean values of the facial expression intensity of all scenes for each intensity level. Also, these values are divided by the number of facial feature values for comparison. For example, in "happiness1" in the previous method, the value $\frac{E}{11}$ is shown in Table 3 because eleven facial feature values are used to calculate the facial expression intensity of the previous method. Likewise, in the proposed method, the value $\frac{E'}{2}$ is shown because the proposed method uses two facial feature values as shown in Table 1. Table 3 shows the mean intensity values of anger, disgust, and fear as well as happiness and surprise. Note that we do not define the intensity levels for anger, disgust, and fear because there are few scenes in which the subjects express these facial expressions.

4.2.2 MMI Datasets

Tables 4, 5 and 6 show the facial expression intensity in each facial expression of each subject. "Nonemotional" represents the mean value of the facial expression intensity of expressionless frame, while "emotional" represents the mean value of the facial expression intensity of the frame in which the expression is completely exposed.

5 Consideration

5.1 Lifelog Videos

In the previous method, subjectB was the only subject with agreement between the facial expression intensity levels and estimated facial expression intensity. On the other hand, in the proposed method, in the expression intensity of "happiness," $happiness3 > happiness2 > happiness1$ were obtained in subjects A, B, D, and F. In these subjects, it was confirmed that the estimated facial expression intensity increases according to the intensity levels. In addition, it is considered that the result in the proposed method is better because the facial expression intensity is larger than the previous method. In other facial expressions, some of the expression intensities in the previous method are larger. These results seem to be caused by the feature selection and standardization of the feature values. About "happiness," the t-test was carried out for each method with happiness1 and happiness2, happiness2 and happiness3, and happiness1 and happiness3. This result is shown in Table 7. From Table 7, there was no significant difference in the previous method, whereas in the proposed method there were significant differences between happiness2 and happiness3, and happiness1 and happiness3. In the proposed method, it is considered that the reason

Table 3 Mean values of estimated facial expression intensity

Subject	Method	Happiness1	Happiness2	Happiness3	Surprise1	Surprise2	Anger	Disgust	Fear
A	Previous	0.016	0.024	0.013	–	–	–	–	–
	Proposed	0.215	0.253	0.441	–	–	–	–	–
B	Previous	0.007	0.011	0.025	–	–	0.042	0.036	–
	Proposed	0.484	1.752	1.956	–	–	0.055	−0.058	–
C	Previous	0.014	8.571×10^{-4}	−0.031	0.018	0.053	–	0.031	0.098
	Proposed	0.856	0.524	0.645	0.223	0.045	–	0.046	−0.031
D	Previous	−0.005	−0.036	−0.003	0.028	–	–	–	–
	Proposed	0.266	0.341	0.462	0.019	–	–	–	–
E	Previous	0.012	0.023	−0.058	0.057	0.041	–	–	–
	Proposed	0.056	0.358	0.019	0.279	0.034	–	–	–
F	Previous	0.004	−0.001	0.019	0.042	0.04	–	–	0.089
	Proposed	0.620	0.670	1.136	−0.091	0.360	–	–	−0.170

Table 4 Facial expression intensity for anger

Subjects	Proposed method		Previous method	
	Nonemotional	Emotional	Nonemotional	Emotional
Id1809	−0.594	0.406	−0.045	0.066
Id1866	−1.140	0.394	0.010	−0.009
Id1874	−0.587	0.055	0.011	0.002
Id1900	−0.312	0.196	0.003	0.002
Id1973	−1.394	0.187	0.018	−0.008

Table 5 Facial expression intensity for disgust

Subjects	Proposed method		Previous method	
	Nonemotional	Emotional	Nonemotional	Emotional
Id1811	−0.597	0.484	−0.008	0.014
Id1821	−0.847	0.447	−0.013	0.011
Id1832	−0.136	0.423	−0.007	0.004
Id1875	−0.755	0.536	−0.038	0.032
Id1882	−0.318	0.046	−0.011	0.008
Id1952	0.111	−0.030	−0.068	0.066
Id1964	−0.774	0.327	−0.007	0.012
Id1991	−0.212	0.178	−0.010	0.006

Table 6 Facial expression intensity for sadness

Subjects	Proposed method		Previous method	
	Nonemotional	Emotional	Nonemotional	Emotional
Id1804	−0.629	0.723	0.003	0.0002
Id1827	0.003	0.095	0.005	−0.008
Id1863	0.348	−0.020	0.003	−0.002
Id1943	0.221	−0.149	0.006	−0.006

Table 7 P and T values of T-Test for lifelog videos

Facial expression	$P(T \leq t)$		t	
	Previous	Proposed	Previous	Proposed
Happiness1 and 2	0.619	0.466	−0.50	−0.72
Happiness2 and 3	0.414	2.64×10^{-8}	−0.82	−5.62
Happiness1 and 3	0.266	3.31×10^{-9}	−1.11	−6.15

Table 8 P and T values T-Test for MMI

Facial expression	$P(T \leq t)$		t	
	Previous	Proposed	Previous	Proposed
Anger	0.0181	1.43×10^{-18}	−1.35	−11.51
Disgust	2.03×10^{-43}	5.73×10^{-31}	−16.34	−12.98
Sadness	9.15×10^{-7}	0.445	5.07	−0.76

why there is no significant difference between happiness1 and happiness2 was that the difference of the facial expression between them is very small.

5.2 *MMI Datasets*

About "anger" in the previous method, the facial expression intensity of "emotional" of most subjects is lower than that of "nonemotional." On the other hand, in the proposed method, this point is improved and it is considered that the expression intensity is estimated correctly. Also "sadness" in the previous method, the expression intensity of "emotional" of all subject is lower than that of "nonemotional." On the other hand, in the proposed method, half of subjects can be estimated correctly. It is considered to be difficult to estimate the facial expression intensity because the difference of the facial expression of subjects is very small. About "disgust," compared to other facial expressions, the facial expression intensity is estimated correctly. However, facial expression intensity of id1952 is smaller in "emotional" than in "nonemotional." This subject tilts his face. This degrades the precision of the facial expression intensity. In each facial expression, we performed the t-test with "emotional" and "nonemotional." This result is shown in Table 8.

Significant differences were found in all facial expressions in the previous method. Half of the subjects are identified by "anger" as shown in Table 4. In "disgust," compared to other facial expression, the estimation accuracy was good as shown in Table 5. In "sadness," the previous method fails to correctly estimate the facial expression intensity because the t value is positive. On the other hand, in the proposed method, significant differences were found in "anger" and "disgust." The reason why no significant difference was found in "sadness" is considered to be attributable to half of the subjects not being correctly estimated as shown in Table 6.

6 Conclusion

In this paper, we proposed new facial feature values and a new facial expression intensity based on them. We estimated facial expression intensity for lifelog videos, and compared the proposed method with the previous method. As a result, it was confirmed that the accuracy of the proposed method was higher for "happiness" than that of the previous method. In the MMI data set, it was confirmed that the expression intensity can be correctly estimated for "anger," "disgust," and "sadness."

In this paper, the inclination of the face is not taken into consideration. In the lifelog videos, since it is assumed that the face moves, it is necessary to calculate the feature values taking into consideration the inclination of the face. It is also necessary to increase the number of lifelog videos and perform sufficient experiments.

An emotional scene can be detected from a lifelog video as the sequence of the frames having high facial expression intensity values. There is, however, no explicit definition to find the frame sequence. The method to appropriately find emotional scenes will be developed in the future work.

Whether there was a positive or negative correlation between the facial feature value and the facial expression intensity was subjectively determined in the proposed method. Since this will be ineffective, we intend to automatically estimate the correlation by using a statistical analysis. In addition, automatically selecting appropriate training data set is included in the future work.

Acknowledgements This research is supported by Japan Society for the Promotion of Science, Grant-in-Aid for Young Scientists (B), 15K15993.

References

1. Gemmell, J., Bell, G., Luederand, R., Drucker, S., Wong, C.: MyLifeBits: fulfilling the Memex vision. In: Proceedings of the 10th ACM International Conference on Multimedia, pp. 235–238 (2002)
2. Aizawa, K.: Information Processing of Experience—Acquisition and Processing of Life Log. Technical research report of the Institute of Electronics, Information and Communication Engineers, Pattern recognition/media understanding, vol. 103, no. 738, pp. 1–9 (2004) (in Japanese)
3. Aizawa, K., Hori, T., Kawasaki, S., Ishikawa, T.: Capture and efficient retrieval of life log. In: Pervasive 2004 Workshop on Memory and Sharing Experiences, pp. 15–20 (2004)
4. Nomiya, H., Hochin, T.: Unsupervised emotional scene detection from lifelog videos using cluster ensembles. Int. J. Softw. Innov. **1**(4), 1–15 (2013)
5. Nomiya, H., Morikuni, A., Hochin, T.: An emotional scene retrieval framework for lifelog videos using ensemble clustering. Int. J. Softw. Innov. **3**(3), 1–13 (2015)
6. Nomiya, H., Hochin, T.: Emotional scene retrieval from lifelog videos using evolutionary feature creation. Stud. Comput. Intell. **612**, 61–75 (2015)
7. Morikuni, A., Nomiya, H., Hochin, T.: Expression strength for the emotional scene detection from lifelog videos. Int. J. Comput. Inf. Sci. **16**(1), 32–39 (2015)
8. Luxand, Inc.: Luxand FaceSDK 4.0. http://www.luxand.com/facesdk
9. Sakaue, S., Hochin, T., Nomiya, H.: Estimation of expression intensity using facial feature values by positional change of facial feature points. J. Inf. Process. Soc. Japan Kansai Branch (2016) (in Japanese)
10. MMI Facial Expression Database. http://mmifacedb.eu/
11. Pantic, M., Valstar, M.F., Rademaker, R., Maat, L.: Web-based database for facial expression analysis. In: Proceedings of IEEE International Conference on Multimedia and Expo, pp. 317–321 (2005)

Developing a Framework to Support Designing of Active Learning Class

Isao Kikukawa, Chise Aritomi, Shoichi Nakamura and Youzou Miyadera

Abstract This study aims to provide teachers who are contemplating to adopt active learning in existing classes with a framework that facilitates converting existing classes to active learning ones and to realize situation where teachers can instantly convert to active learning. Therefore, we develop "Four-quadrant Active Learning Designing Method" as an idea for "framework to support designing of active learning class". It is considered that use of "Four-quadrant Active Learning Designing Method" further facilitates converting existing classes to active learning ones for teachers. Furthermore, in this study, we examine effectiveness of "Four-quadrant Active Learning Designing Method" through classroom designing and practice based on it.

Keywords Active learning · Instructional design · Framework · Designing support · Classroom practice

1 Introduction

Recently classroom practice adopting "Active Learning (AL)" is in need in many educational institutions including universities. Many researches have been conducted researches in designing method for active learning classes and published

I. Kikukawa (✉) · C. Aritomi
Tokoha University, 325, Obuchi, Fuji-shi, Shizuoka 417-0801, Japan
e-mail: kikukawa@fj.tokoha-u.ac.jp

C. Aritomi
e-mail: aritomi@fj.tokoha-u.ac.jp

S. Nakamura
Fukushima University, 1, Kanayagawa, Fukushima-shi, Fukushima 960-1296, Japan
e-mail: nakamura@sss.fukushima-u.ac.jp

Y. Miyadera
Tokyo Gakugei University, 4-1-1, Nukuikita-machi, Koganei-shi, Tokyo 184-8501, Japan
e-mail: miyadera@u-gakugei.ac.jp

R. Lee (ed.), *Software Engineering, Artificial Intelligence, Networking and Parallel/Distributed Computing*, Studies in Computational Intelligence 721,
DOI 10.1007/978-3-319-62048-0_10

them as literature (e.g., [1–8]). General framework and designing process are discussed in such literature, and case studies of classroom practice based on the theories are referred to. Therefore such literature is a quite useful source of information in designing active learning classes. However, the contents of such literature are not always easy for teachers who are contemplating to adopt active learning in existing classroom (mainly university teachers are supposed in this study). We feel some kind of "framework to support designing active learning class" is necessary in order for teachers to make an immediate switch to active learning classroom. So this study develops "Four-quadrant AL Designing Method" as an idea for "framework to support designing of active learning class".

In this paper, Chap. 2 describes outline of active learning classes to extract intrinsic problems of class designing among teachers. Chapter 3 organizes factors to solve these problems and explains "Four-quadrant AL Designing Method" which satisfies the factors in detail. Chapter 4 shows case studies of improved classes by using "Four-quadrant AL Designing Method". Chapter 5 describes classroom practice based on class designing in Chap. 4, followed by examination of implemented active learning classes through questionnaire survey, etc. Chapter 6 discusses effectiveness of "Four-quadrant AL Designing Method" based on the results of Chaps. 4 and 5.

2 Problems in Designing Active Learning Class

2.1 *Outline of Active Learning Class*

Multiple definitions for "Active Learning" exist, and the leading one is advocated by Bonwell and Eison [1]. Bonwell and others define "Active Learning" as "anything that involves students in doing things and thinking about the things they are doing" [9]. Furthermore, Fink developed the definition by Bonwell and others to present "A Holistic View of Active Learning (Fig. 1)" [1]. Figure 1 indicates that it is possible to recognize "Active Learning" as trinity learning style of "Information and Ideas", "Experiences" and "Reflecting".

In this study, "Active Learning" is interpreted based on definition by Fink among many views that cover a broad range. It is because (1) the view of Fink is considered to be the sufficient condition to execute active learning, (2) it also presents "Integrated Course Design Model", a methodology in designing active learning class.

Table 1 shows "Integrated Course Design Model". In this model, teachers are expected to plan active learning lessons through the following Initial Phase, Intermediate Phase and Final Phase. First, as is shown in Fig. 2, the components of "Situational Factors", "Learning Goals", "Feedback & Assessment" and "Teaching & Learning Activities", the most important components in course designing, are analyzed and integrated in Steps 1 to 5 so that they keep organic connection (Initial

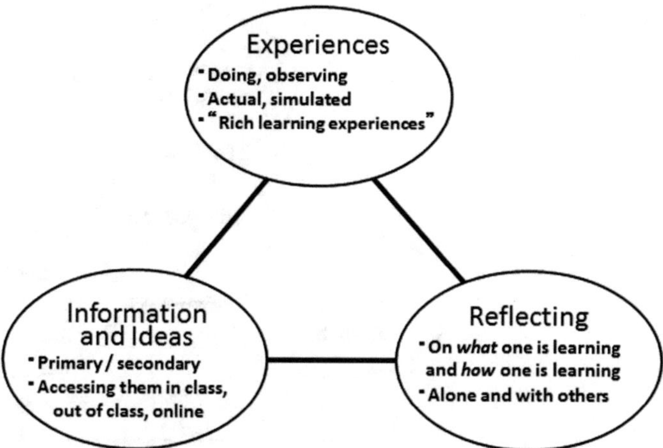

Fig. 1 A holistic view of active learning (*Source* Fink 2013)

Table 1 Steps in integrated course design (*Source* Fink 2013)

Initial Phase: *Build Strong Primary Components*
1. Identify important situational factors.
2. Identify important learning goals.
3. Formulate appropriate feedback and assessment procedures.
4. Select effective teaching and learning activities.
5. Make sure the primary components are integrated.
Intermediate Phase: *Assemble the Components into a Coherent Whole*
6. Create a thematic structure for the course.
7. Select or create a teaching strategy.
8. Integrate the course structure and the instructional strategy to create an overall scheme of learning activities.
Final Phase: *Finish Important Remaining Tasks*
9. Develop the grading system.
10. Debug the possible problems.
11. Write the course syllabus.
12. Plan an evaluation of the course and of your teaching.

Phase). Next in Steps 6 to 8, course structure is examined and constructed based on the analysis in the Initial Phase while key topics of the course are extracted. Overall scheme of learning activities is modulated and determined after "Teaching Strategy" for which each key topic is planned (Intermediate Phase). Then in Steps 9 to 12, decision of grading rules, prediction of possible problems and precautions, making of course syllabus, and plan of course evaluation are implemented (Final Phase).

In this study, any classroom lesson that can be designed in the framework or steps shown above is considered "Active Learning Class".

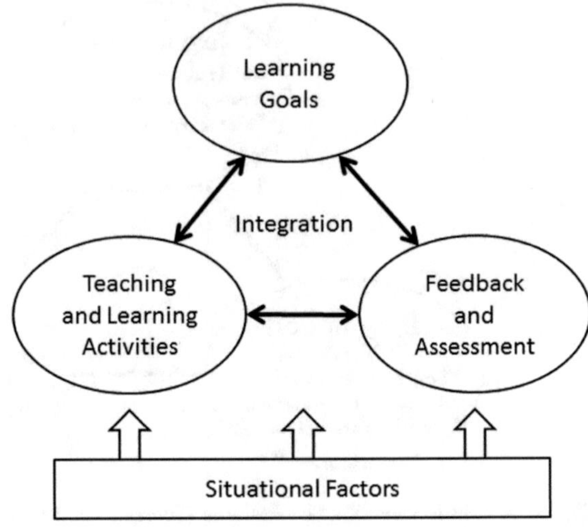

Fig. 2 Key components in integrated course design (*Source* Fink 2013)

2.2 Problems in Converting Existing Class Sessions to Active Learning

This study aims to provide teachers who try to adopt active learning in existing class sessions with a framework which enables them to "convert to active learning class" more smoothly. Therefore problems in converting existing class sessions to active learning ones are extracted with the preceding section in mind. Specifically, we will implement review of existing classes following "Integrated Course Design Model" process by Fink. Through the review, "steps which are most difficult for teachers to implement" are discussed, from which problems to be solved are put in order.

First in Steps 1 to 5 (Initial Phase), initial designing of "Situational Factors", "Learning Goals", "Feedback & Assessment", "Teaching & Learning Activities" and so on are conducted based on existing class sessions. It was judged that these were not the most difficult steps for implementation since it was possible to make use of existing classroom information expansively. Similarly, it was judged that Steps 9 to 12 (Final Phase) and Step 6 were not the most difficult steps for implementation, since it was possible to design these based on existing class sessions. The result showed that Step 7 was found to be the most difficult one to implement. The reason is because following problems were found; it is difficult for teachers to establish examination policy of contents such as how learning activities in existing classroom can be improved and new "Teaching Strategies" can be planned; it is difficult to judge if "Teaching Strategies" after examination is appropriate or not, when "Teaching Strategies" need to be built in Step 7 by effectively combining various learning activities such as activities within the classroom as well as outside classroom ones, and individual and group activities [1]. Step 8 can be achieved by comprehensively summarizing the results of Step 7. Therefore Step 8 will not be a difficult step for implementation once the difficulty of Step 7 is solved. "Problems in converting existing classes to Active Learning" in this discussion are indicated below.

- Problem 1: It is difficult to establish examination policy of contents such as how learning activities in existing classrooms can be improved and new "Teaching Strategies" can be planned.
- Problem 2: It is difficult to judge if "Teaching Strategies" after examination are appropriate or not.

3 Four-Quadrant AL Designing Method

The topic of discussion in this chapter is the requirements based on the previous chapter in order to solve the problems in converting existing classes to active learning ones, and we develop "Four-quadrant AL Designing Method" which satisfies these requirements. "Four-quadrant AL Designing Method" is considered to solve the problems and achievement of the aim of this study will become possible.

3.1 Requirements to Solve the Problems

Requirements to solve the problems described in Chap. 2 are examined. Specifically, problem solving was attempted from the viewpoint of "What are the requirements to facilitate converting existing classroom lessons to active learning?". The results are shown below.

- Requirement 1: Steps to improve existing learning activity are indicated (responds to Problem 1). These steps for improvement are supposed be equipped with sufficient flexibility so that they can accommodate "prototype classroom designing" and "class improving designing after class implementation" according to the needs of teachers.
- Requirement 2: Clear standpoint to examine learning activity improvement (responds to Problems 1 and 2).
- Requirement 3: Existence of rules to judge whether the improvement results are appropriate or not (responds to Problem 2).

3.2 Approach to Satisfy the Requirements

This section describes the approach to satisfy the requirements indicated in the previous section.

First, in order to meet Requirement 1, designing results of "Teaching Strategies" will be presented in "Castle Top Diagram" by Fink and designing process of Castle Top Diagram newly becomes a model. This process is constructed to be repetitive with interaction of processes instead of one-way processes so that sufficient flexibility is guaranteed. Next, "individual activity in class", "group activity in class", "individual activity out of class" and "group activity out of class" are established as points of view to meet Requirement 2, and Castle Top Diagram is expanded so that these activities are combined in well balance for "Teaching Strategies" (hereafter called Expanded Castle Top Diagram). Furthermore in order to meet Requirement 3, information to judge "whether the improved results are appropriate or not" are indicated as a guideline in designing process model to be constructed. The newly constructed model based on these approaches will be described in the following section. This new model is called "Four-quadrant AL Designing Method" in this study.

3.3 Development of "Four-Quadrant AL Designing Method"

In this section, "Four-quadrant AL Designing Method" is developed as "framework to support designing active learning class".

First, Castle Top Diagram of Fink was expanded. The original Castle Top Diagram is made to clarify combination and sequence of activities in and out of class while referring to learning activities divided into two categories of "activities in class" and "activities out of class" (Fig. 3) [1]. In this study, it was expanded so that it can respond to four categories of "individual activity in class", "group activity in class", "individual activity out of class" and "group activity out of class". Each learning activity is organized in four-quadrant matrix so that it is easy to confirm which learning activity belongs to the aforementioned four categories (Fig. 4) (in creating Fig. 4, [10] was referred to). The contents of Fig. 4 can be modified to meet the needs of individual teachers. Figure 5 is template of Expanded

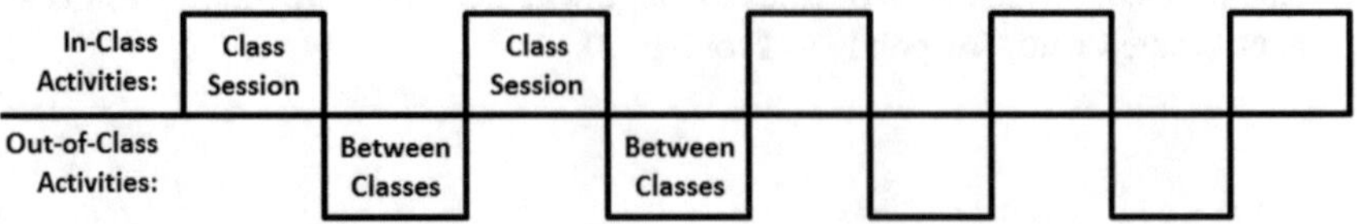

Fig. 3 Template of original castle top diagram (*Source* Fink 2013)

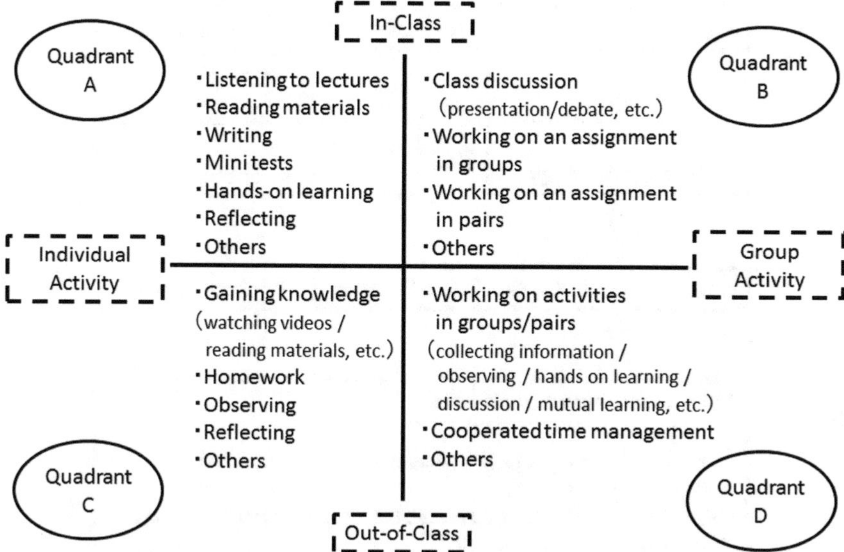

Fig. 4 Four-quadrant matrix of learning activities

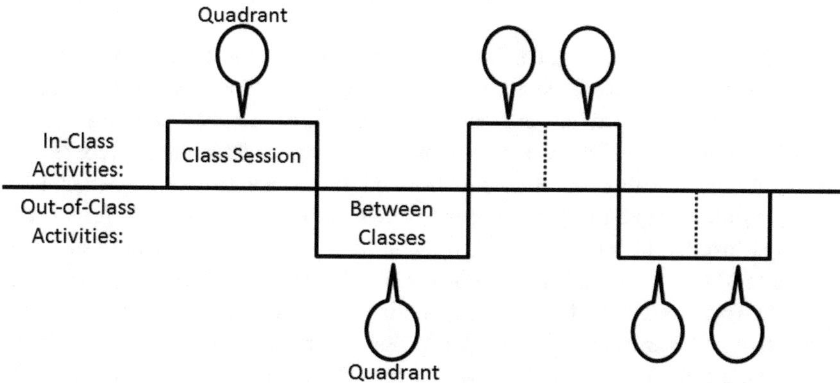

Fig. 5 Template of expanded castle top diagram

Castle Top Diagram. Teachers can clarify combination and sequence of activities in and out of class while referring to the contents of Fig. 4, and in addition to that, teachers can clarify combination and sequence of individual activity and group activity. Contents of Expanded Castle Top Diagram gained as a result of work so far become the new "Teaching Strategies". Span of "Teaching Strategies" is assumed to be "one to three weeks" as was indicated by Fink.

Next, process model for this designing method was developed (Fig. 6). In this process model, the number of class sessions is examined first where template shown

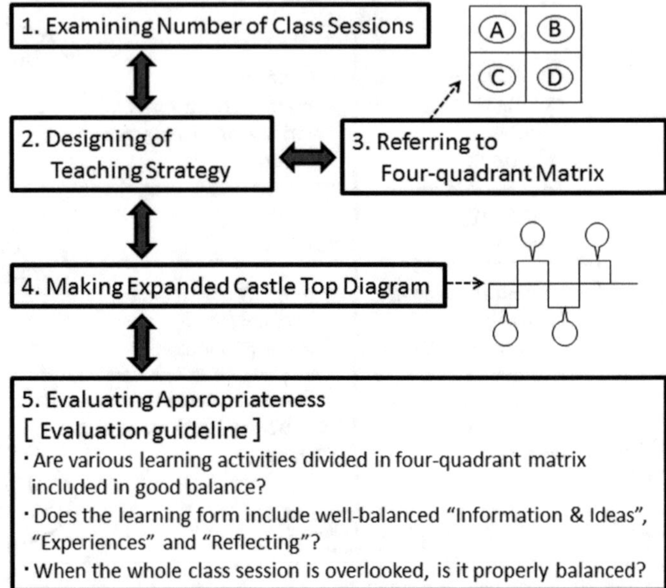

Fig. 6 Production step of expanded castle top diagram

in Fig. 5 is applied (1 of Fig. 6). Designing of "Teaching Strategies" starts after the number of class sessions is determined (2 of Fig. 6). In doing so, combination and sequence of various activities are clarified (2 of Fig. 6) in reference to the learning activities (3 of Fig. 6) in the above-mentioned four-quadrant matrix (contents of Fig. 4). When learning activities are inserted in the Castle Top Diagram (4 of Fig. 6), symbol of quadrant to which the activity belongs (one of A, B, C, and D in Fig. 4) is inserted as well (4 of Fig. 6). The final step is the evaluation of appropriateness of Castle Top Diagram (5 of Fig. 6). Teachers check the appropriateness of Castle Top Diagram following the guideline of 5 of Fig. 6. Specifically, check is done from the viewpoint of "are the various learning activities divided in the four-quadrant matrix included in good balance?", etc. If any trouble is found after checking, go back the steps in Fig. 6 to modify Castle Top Diagram. Finished version of Expanded Castle Top Diagram is made by repeating this procedure as much as necessary. This process model is positioned as a part of iterative version of "Integrated Course Design Model" as is shown in Fig. 7. That is to say, judgment by overlooking the whole class session is available in evaluating appropriateness of the Designed Castle Top Diagram. Also using a cycle like Fig. 7 allows flexible response to "prototype class design" and "class improvement design after implementation", etc.

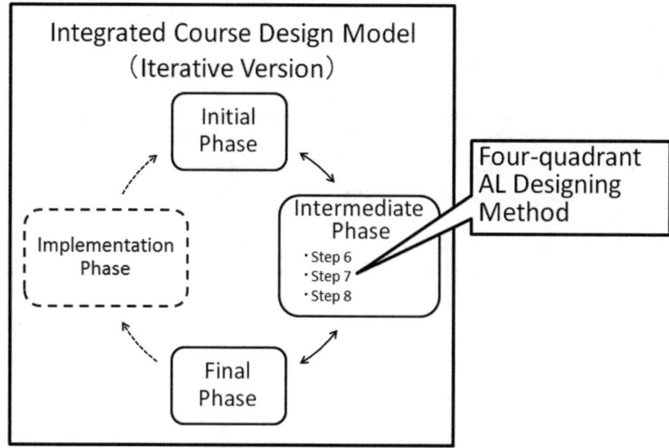

Fig. 7 Image of iterative version of integrated course design model

4 Case of Improved Class Using "Four-Quadrant AL Designing Method"

In this chapter, "Database" class in active learning style conducted by one of the authors is described as a class improvement case using "Four-quadrant AL Designing Method". The purpose of "Database" class is to mainly master the basics of SQL (including operation skills on actual computers).

Figure 8 shows "Teaching Strategy" before active learning. "Teaching Strategy" in Fig. 8 is a structure designed for one week. After listening to a lecture delivered by the teacher, students work on worksheets concerning contents of the lecture. Their answers are checked by the teacher upon completion of the worksheet, and they operate PC (carry out SQL) based on the contents of the worksheet, and validate the operation of worksheet. They are also instructed to individually review what is learned after the class. Classes based on such "Teaching Strategy" were conducted total 15 times. The weakness of this "Teaching Strategy" was that "it was difficult to complete all these activities in one class session because each individual activity in class was time consuming". The students were instructed to individually work as homework on the contents that were not finished during the class.

When we confirmed to see which quadrant in Fig. 4 each activity in Fig. 8 belonged to, we found activities in quadrant "A" were dominant with some activities in quadrant "C" were included. So new "Teaching Strategy" was designed by allocating class time (class sessions) for two weeks and adding activities in quadrant "B" and "D", as well as complementing activities in quadrant "C". The result is shown in Fig. 9.

Figure 9 shows two class sessions and flow of activities in between. Here, "gaining knowledge" which was delivered in lecture during class session is changed

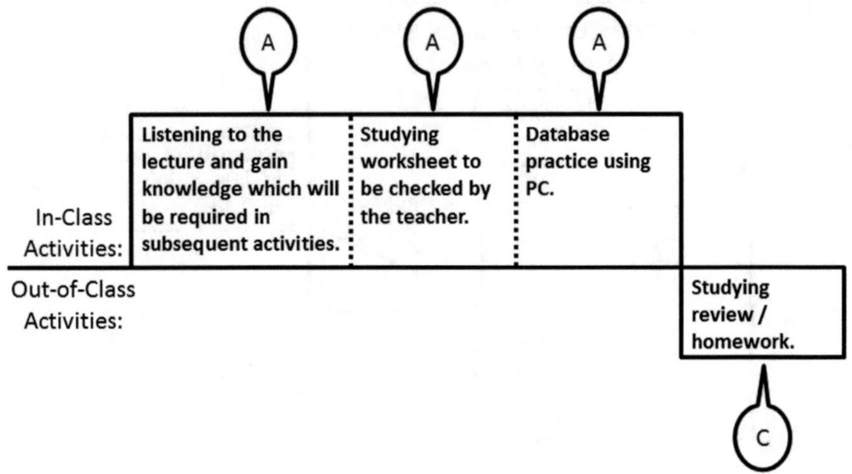

Fig. 8 Teaching strategy before improvement

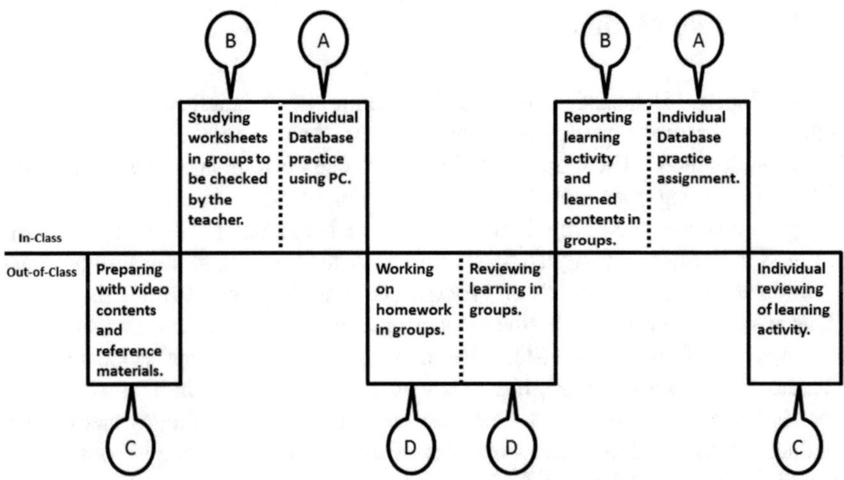

Fig. 9 Teaching strategy after converting to active learning

to an activity of viewing online video contents and reference materials as pre-class activity. This allowed increased time for operating database on actual computers in addition to more time to work on worksheet during class session. During the class session after preparation, students worked on a worksheet as a group followed by an individual database operation on actual computer. Activity to work on a worksheet for review was prepared as homework for "between this class and next the class". This worksheet for review was formatted in problem solving in extracurricular group work, just like the learning activity in class. During the second class session,

Table 2 Class structure after converting to active learning

Phase	Major topics
1 (3 weeks)	Introduction to database, SQL overview, etc.
2 (2 weeks)	INSERT statement, SELECT statement, etc.
3 (2 weeks)	Using WHERE clause, Sorting
4 (2 weeks)	UPDATE statement, DELETE statement, etc.
5 (2 weeks)	Joining tables (1): INNER JOIN
6 (2 weeks)	Joining tables (2): LEFT JOIN, RIGHT JOIN
7 (2 weeks)	Basics of database design

activities results and contents of group activities between the classes were reported so that students could individually work on hands-on assignments of database operations on actual computers. After the class session, students were required to review their learning activity and write a comment as a summary. When each activity in Fig. 9 was checked to see which quadrant in Fig. 4 it would fall in, it was confirmed that activities in all the quadrants were included in good balance.

Whole class session was examined based on the new "Teaching Strategy" that was designed in the abovementioned process. Since there were two class sessions in the new "Teaching Strategy", they are considered as "a phase". The whole course (total 15 times) was reconstructed, altering learning contents and order of learning, so that it would consist of multiple "phases". The result is shown in Table 2. Also video contents and worksheets were made so that class sessions based on Table 2 and Fig. 9 can be implemented.

5 Class Implementation and Results

"Database" course based on the result of class improvement described in Chap. 4 was conducted in the fall semester of 2015–2016 school year at Fuji Campus, Tokoha University. In its implementation, Dynamic LMS [11] developed by the authors in order to support individual review activities and group activities, etc., was used. In learning environment, dynamic group changes and role changes are available in addition to smooth learning progress. Therefore an operation method of Dynamic LMS which includes the viewing method of video contents and the comment input method in reviewing learning activities was added in Phase 1 in Table 2 on top of outline of active learning class and basic operation of database. Because of this, class sessions by "Teaching Strategy" indicated in Fig. 9 was moved to be implemented after Phase 1. In this class, written exam (perfect score is 100) is given at the end of semesters every year, and the written exam which had the same contents as the one for the school year 2014–2015 was given at the 16th class in the school year 2015–2016.

Based on the above, here is consideration concerning class implementation from the results of written exam given at the end of term of "Database" and student survey. First, exam results were compared between 2014–2015 school year (before improvement, 34 students) and 2015–2016 school year (after converting to active learning, 36 students on Tuesday, 4th period, and 46 students on Thursday, 3rd period). Average score in 2014 was 72.74 (SD:15.99) while Tuesday class in 2015 scored 82.31 (SD:17.37), confirming improved results from 2014. Significant difference was examined in t-test (independent measures), which was confirmed by 5%. Average score of Thursday class (77.22 (SD:17.09)) did show improvement from that of 2014–2015, however, no significant difference was observed. Next is survey result. Survey topics were designed to respond to four quadrants (in five ranks with 5 the highest). Almost all topics (for example, "did your individual activities before the classes help you in your learning?", "did your group activities in the classes help you in your learning?", etc.) scored relatively good average of 3.5–4.1 points in both Tuesday and Thursday classes in the class implementation. Also such comments as these were written as free writing answers for questions; "Structure of learning outside classroom was easy to understand and was easy to deal with", "dealing with questions in groups raised motivation of individuals", "There were opportunities to do preparation and review, and it was easy to understand them".

Comprehensive consideration from the above results reveals that "Database" in 2015–2016 school year effectively adopted "Active Learning" in comparison with 2014–2015 school year, which resulted in higher exam scores.

6 Consideration for "Four-Quadrant AL Designing Method"

Effectiveness of "Four-quadrant AL Designing Method" is discussed in this chapter based on the results of Chaps. 4 and 5.

As was said in Chap. 5, conversion to active learning in "Database" class by "Four-quadrant AL Designing Method" seems to be overall successful. When considering factor for the success from the point that Fink viewed "Teaching Strategy" as important in order to realize effective class, it was because of analyzing existing "Teaching Strategy" from the four events of "individual activity in class", "group activity in class", "individual activity out of class", and "group activity out of class", which enabled designing new "Teaching Strategy" by supplementing activities in insufficient event(s). Furthermore, the whole class session was reconstructed based on the new "Teaching Strategy" in "Database". That is to say, one of the factors we can point out is that the effectiveness brought by the cycle shown in Fig. 7 functioned fully. From these, it is considered that "Four-quadrant AL Designing Method" has a certain degree of effectiveness. Also this designing method can be positioned as a framework for teachers to instantly convert to active learning classes when compared with related researches. However, we do not have enough data to consider the conclusion at this point yet concerning "solution for problems described in Sect. 2.2". Therefore in the future, we would like to ask teachers other than the author to make use of this designing method, and conduct more specific evaluation concerning this designing method by surveys to such teachers. Also "Database" is in progress in the school year 2016–2017 with slight improvement on that of the school year 2015–2016. We would like the results of this implementation to help deeper consideration for this designing method.

7 Conclusion

This paper is a report of class improvement and class implementation based on "Four-quadrant AL Designing Method" which was developed as an idea of "framework to support designing of active learning class". Certain level of effectiveness of this designing method is confirmed from consideration through the results of classroom implementation. Future issues are collecting as many usage cases of "Four-quadrant AL Designing Method" as we can, and more specific evaluation and consideration concerning this designing method through surveys to teachers who use this designing method.

References

1. Fink, L.D.: Creating Significant Learning Experiences: An Integrated Approach to Designing College Courses. Wiley (2013)
2. Gagnon, G.W., Collay, M.: Constructivist Learning Design: Key Questions for Teaching to Standards. Corwin Press (2005)
3. Bergmann, J., Sams, A.: Flip Your Classroom: Reach Every Student in Every Class Every Day. International Society for Technology in Education (2012)
4. Bergmann, J., Sams, A.: Flipped Learning: Gateway to Student Engagement. International Society for Technology in Education (2014)
5. McCain, D.V., Tobey, D.D.: Facilitation Basics. American Society for Training and Development (2004)
6. Reigeluth, C.M., Carr-Chelman, A.A.: Instructional-Design Theories and Models, Volume III: Building a Common Knowledge Base. Routledge (2009)
7. Barkley, E.F., Cross, K.P., Major, C.H.: Collaborative Learning Techniques: A Handbook for College Faculty. Wiley (2014)
8. Ambrose, S.A., Bridges, M.W., DiPietro, M., Lovett, M.C., Norman, M.K.: How Learning Works: Seven Research-Based Principles for Smart Teaching. Wiley (2010)
9. Bonwell, C.C., Eison, J.A.: Active Learning: Creating Excitement in the Classroom. ASHE-ERIC Higher Education Report No. 1. The George Washington University, Washington, D.C. (1991)
10. Mizokami, S.: Active Learning to Kyoju Gakushu Paradigm no Tenkan. Toshindo, Tokyo (2014) (in Japanese)
11. Kikukawa, I., Aritomi, C., Miyadera, Y.: Development of a LMS with dynamic support functions for active learning. In: Computer and Information Science, pp. 103–117. Springer International Publishing (2016)

Where Do Drivers Look When Driving in a Foreign Country?

Yumiko Shinohara and Yukiko Nishizaki

Abstract The importance of developing automated driving systems based on driver characteristics is increasing rapidly. Although some previous studies indicated differences in driver eye movements caused by driving experience or the situation, few studies have focused on the effects of familiarity with the driving environment on driver eye movements. In this study, we investigated differences in eye movements, especially fixation duration and location, between novice and expert drivers when driving abroad. The eye movements of about 50 participants were measured during driving simulations in two situations: driving in their own country and in a foreign one. Results indicated that experienced drivers were more likely to be influenced by familiarity of the driving situation. This results show the need to develop an automated driving system that considers drivers' driving background.

Keywords Eye movements · Eye tracking · Driving experience · Driving in a foreign country · Fixation duration

1 Introduction

The number of studies on the development of automated driving vehicles is increasing rapidly with the burgeoning technologies. Although a car accident may be attributable to the driver, vehicle, or environment, about 90% of critical reasons for accidents have to do with the drivers [1, 2]. Moreover, about 41% of driver-related critical reasons have been identified as recognition errors [1]. Thus, a major purpose

Y. Shinohara (✉)
Graduate School of Information Science, Kyoto Institute of Technology, Kyoto, Japan
e-mail: yumiko.shinohara.918@gmail.com

Y. Nishizaki (✉)
Information and Human Sciences, Kyoto Institute of Technology, Kyoto, Japan
e-mail: yukikon@kit.ac.jp

R. Lee (ed.), *Software Engineering, Artificial Intelligence, Networking and Parallel/Distributed Computing*, Studies in Computational Intelligence 721,
DOI 10.1007/978-3-319-62048-0_11

in the development of automated driving vehicles is to support safer driving and to reduce car accidents caused by human errors, especially human recognition errors. However, the realization of autonomous vehicles will occur in phases. We need to work with automated driving systems during step-wise progression through levels from "non-automation" to "full automation" [3]. To realize effective cooperation between a driver and an automated driving system, operation of the driving system should depend on the driver's characteristics and the driving situation.

This study was conducted to develop an automated driving system that considers driver characteristics and the driving situation. In particular, we focused on personal differences in perspective recognition, one of the main driver-related factors in car accidents. Some previous studies have shown that eye fixation duration of novice drivers is longer than that of expert drivers and that the fixation location differs between novice and expert drivers [4–6]. Novice drivers tend to focus on a guard rail to determine their vehicles' positions [4]. When turning at a curve, expert drivers gaze to the inside edge of the lane, whereas novice drivers focus on the area from the center to the outside edge of the lane. Fixation duration is also affected by the spatial extent of the useful field of view, i.e., the visual field where visual information can be extracted in a single fixation without eye or head movements [7]. Increased cognitive workload in a complex traffic environment causes a narrowing of the useful field of view and makes fixation duration shorter [8]. Although many previous studies have revealed differences in eye movements based on driving experience and/or traffic situation, such as the type of road or the amount of traffic information [7–9], few studies have focused on eye movements while driving overseas, where the driving environment, such as types of roads and traffic signs, is quite different.

The primary purpose of this study was to investigate whether eye movements are influenced by driving experience and familiarity with the driving environment. Eye movements during driving simulations in the driver's own country and in another country were measured. Analysis of eye movements showed that the fixation duration of expert drivers tended to be shorter than that of novice drivers when driving in a foreign country.

The rest of this paper is structured as follows. Section 2 describes the methods, and Sect. 3 shows the results. Section 4 provides a discussion of the results of the eye tracking and the situational awareness questionnaires. Finally, Sect. 5 concludes the paper.

2 Methods

2.1 Participants

In total, 47 people (23 females, 24 males) between 21 and 59 years of age ($M = 32.3$, $SD = 12.7$) participated. We separated the participants into two groups based on driving experience. Participants who had driven a car for more than 2 years (whole years) and at least once per week were deemed expert drivers (11 females, 11 males). The others were deemed novice drivers (12 females, 13 males).

This study was overseen by Kyoto Institute of Technology's Institutional Review Board. All participants agreed to the treatment and protection of their data.

2.2 *Apparatus and Materials*

We measured eye movements using an eye tracker (X2-60, Tobii, Japan) attached to a 27-inch screen (GW2255, Ben Q Corp., Japan). Participants watched one video for practice (on a highway in Canada) and two videos taken in Osaka, Japan, and San Francisco, USA, 'behind the wheel' (T80, Thrustmaster). Figure 1 shows the experiment room. The video for practice did not show any cultural characteristics, such as types of lane or right- or left-hand traffic. The two videos for measuring eye movements were taken on clear days in October between 1:30 and 3:00 p.m. The videos included right and left turns at least once and some objects, such as a pedestrian, an oncoming vehicle, a preceding vehicle, and a building. The video for practice was ~1 min long, and each of the two videos was 3 min long.

Participants were asked to answer a situational awareness questionnaire after each of the two videos (Tables 1, 2). Each questionnaire included two trick questions so as not to make participants feel as if they should answer "yes" to all questions. The trick questions were Q6(a) and Q6(b) in the questionnaire for the Osaka driving video and Q3(a) and Q3(b) in the questionnaire for the San Francisco driving video. All questions were about the following objects.

A. An object in the right or left lane
B. A pedestrian at a corner
C. A preceding vehicle

Fig. 1 The experiment room

Table 1 The situational awareness questionnaire for the Osaka driving video

No.	Type	Question
Q1	A	Did you see a pale green car parked in the left lane?
Q2	A	Did it have its blinkers on?
Q3	C	Was there a motorcycle ahead of you on the road?
Q4	C	What lane was the motorcycle in?
Q5	C	Which direction did the motorbike turn at the first intersection?
Q6 (a)	Trick	Did you see a dog on the sidewalk?
Q6 (b)	Trick	What side was the dog on?
Q7	E	At the first intersection, if your car wanted to go straight, could it have done so?
Q8	E	Why or why not?
Q9	B	Did you see a pedestrian walking on the left side after the first intersection?
Q10	B	What was the pedestrian doing?
Q11	F	Did you see a store with a big red sign?
Q12	F	What kind of store was it?
Q13	E	What was the speed limit at the location where the video ended?
Q14	D	Did you see a man walking between 2 cars in the street?
Q15	D	Did you see a taxicab pull out in front of your car?
Q16	D	Which direction did the taxicab come from?
Q17	E	At the moment the video ended, what color was the traffic light?

D. A crossing pedestrian or vehicle
E. A traffic sign or signal
F. A focal object

2.3 Procedure

Participants signed a consent form and completed a questionnaire about their driving experience. The experimenter provided a brief introduction and asked the participant to imagine s/he was sitting in the driver's seat of an automated vehicle and to keep holding the wheel during the driving task. Participants watched the two videos taken in Osaka and San Francisco after seeing a short video for practice. The order of these two videos was counterbalanced to control for order effects. The participants were asked to complete the situational awareness questionnaires after each video. The eye tracker was calibrated before each video. While the participant was watching the videos, the experimenter waited outside the room. It took ~ 20 min to complete all procedures.

Table 2 The situational awareness questionnaire for the San Francisco driving video

No.	Type	Question
Q1	A	Did you see a person riding a bicycle at the first intersection?
Q2	A	Where was the bicycle?
Q3(a)	Trick	Did you see a man running at the first intersection?
Q3 (b)	Trick	What direction was he running?
Q4	E	At the first intersection, if your car wanted to turn left, could it have done so?
Q5	E	Why or Why not?
Q6	C	Did the gray car in front of you have its turn signal on?
Q7	C	What direction was it signaling?
Q8	A	Did you see a street car?
Q9	A	What color was it?
Q10	F	Did you see a clock tower?
Q11	F	Which direction did your car turn just before the clock tower?
Q12	D	Did you see a pedestrian wearing a yellow jacket and crossing the street?
Q13	D	What direction was the pedestrian crossing?
Q14	C	Which direction did the double-decker bus turn at the last intersection?
Q15	B	Did you see a pedestrian standing in the street when your car turned left?
Q16	B	Where was he standing?
Q17	A	Did you see a man getting into a car?

2.4 Analysis

To determine where the participants tended to look, we separated a frame into several areas based on the type of road, i.e., straight or curved. During a scene on a straight road, each frame was split into seven parts: the left lane, the right lane, the vehicle's lane, above the middle, buildings on the right, buildings on the left, and in

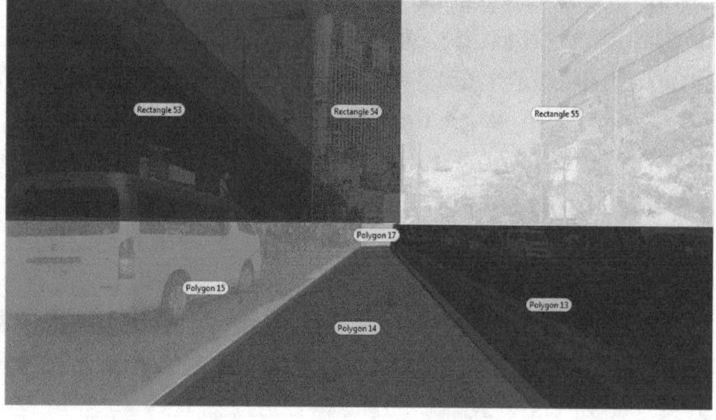

Fig. 2 Example of a separated frame on a straight road

the middle (Fig. 2). When approaching a curve and waiting for an oncoming vehicle, each frame was separated vertically into three areas: the oncoming lane, the middle, and the direction to go. While turning right or left, each frame was divided vertically into three equal parts: the left, the center, and the right. Moreover, when the eyes moved up or down while driving on a curve, each frame was separated horizontally into two areas; the number of areas in this scene was six. In each situation, when there was a specific object, such as a focal object or a pedestrian, the area of that object was also defined. The method for dividing the frames was based on a previous study [9].

3 Results

Although a participant who was classified to expert drivers' group had driven in the US once, we did not remove the data since one day driving experience was not enough to adjust to driving in the US.

3.1 Total Fixation Duration Within the Prescribed Areas

We calculated total fixation duration within the prescribed areas by using Olsson's classification algorithm [10]. Figure 3 shows total fixation duration of experienced and novice drivers, along with the results of two-samples *t*-tests.

The *t*-test on total fixation duration on the area of the left lane during a scene showing a straight road in the Osaka driving video indicated that total fixation duration of expert drivers was significantly greater than that of novice drivers, $t\ (29.6) = 1.95$, $p < 0.05$. In the area of the right lane and the middle area in the same situation, the total fixation duration of novice drivers was significantly longer than those of expert drivers, $t\ (27.5) = 1.75$ and $t\ (26.0) = 2.26$, $p < 0.05$, respectively. When turning right, the *t*-test revealed a marginally significant difference in total fixation duration on the area to the right; expert drivers gazed longer at that area than did novice drivers, $t\,(45) = 1.39, p < 0.10$. When turning left, total fixation duration showed a marginally significant difference between expert and novice drivers, with novice drivers tending to focus longer on the area to the left than expert drivers did, $t\ (29.1) = 1.33$, $p < 0.10$. In the same situation, the *t*-test showed a marginally significant difference in total fixation duration; expert drivers gazed longer at a store with a big red sign than did novice drivers, $t\ (25.4) = 1.64$, $p < 0.10$.

During a scene showing a straight road in the San Francisco driving video, the *t*-test revealed a marginally significant difference in total duration of fixation on the area above the middle of the scene, $t\,(21.1) = 1.41, p < 0.10$. These results showed that expert drivers focused longer on this area than novice drivers did. In the same

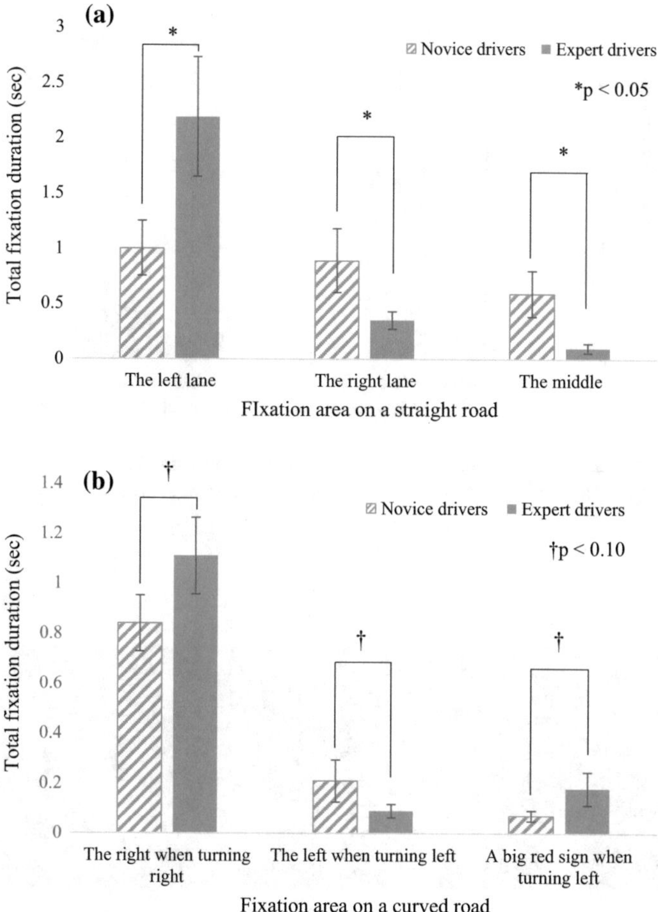

Fig. 3 Novice and expert drivers' mean total fixation duration within each of the prescribed areas in the Osaka driving video: **a** The mean total fixation duration on a straight road, **b** The mean total fixation duration on a curved road (*The error bars represent standard errors*)

situation, expert drivers also exhibited significantly longer total fixation on buildings on the left, $t\ (27.0) = 1.95$, $p < 0.05$ (Fig. 4).

There was no significant difference in fixation on any other area between expert and novice drivers. Figure 5 shows scenarios in which significant differences in total fixation duration appeared.

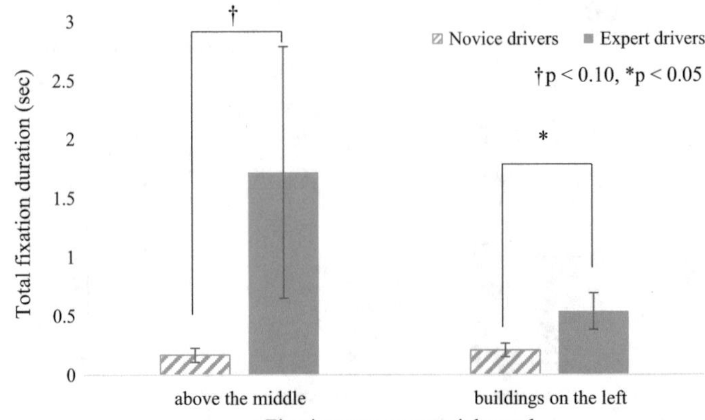

Fig. 4 Novice and expert drivers' mean total fixation duration within each of the presented areas in the San Francisco Driving video. (*The error bars represent standard errors.*)

Fig. 5 Sample frames in each situation where differences were indicated: **a** a straight road in Osaka, **b** a curved road in Osaka (*right*), **c** a curved road in Osaka (*left*), **d** a straight road in San Francisco

3.2 Fixation Duration During the Full Videos Without Divisions into Areas and Scenes

Total fixation duration, not distinguishing among areas, was also calculated for each of the driving videos. Figure 6 shows the mean total fixation duration of expert and novice drivers during the videos. The result of a one-sample *t*-test showed that expert drivers' fixation duration was marginally significantly different between the Osaka driving video and the San Francisco one; it was shorter in the San Francisco driving video than in the Osaka one.

3.3 Situational Awareness Questionnaires

The mean correct answer scores for the seven types of question and the results of a mixed analysis of variance (ANOVA) with one between-subjects variable, namely driving experience (novice vs. experienced), and seven within-subject variables based on the question topics (A to F and trick question), are summarized in Tables 3 and 4.

In the situational awareness questionnaire for the Osaka driving video, the mean rate of correct answers of experienced drivers did not differ from that of novices. A main effect of type of question was seen: F (6, 270) = 63.6, $p < 0.01$. A significant interaction between driving experience and type of question was also apparent: F (6, 270) = 2.89, $p < 0.05$. The mean rate of correct answers for the type A questions about an object in the right or left lane showed a marginally significant difference, with more correct answers given by expert drivers than by novices, F (1, 45) = 3.30, $p < 0.10$. For the type D questions, about a pedestrian at

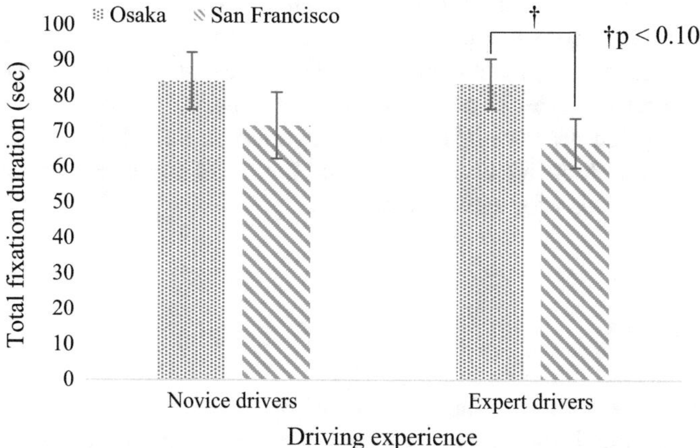

Fig. 6 Novice and expert drivers' mean total fixation duration during the full videos without division into areas and scenes. (*The error bars represent standard errors.*)

Table 3 Results of the situational awareness questionnaire for the Osaka driving video

Type	Novice drivers M(SD)	Expert drivers M(SD)	F
A	0.00 (0.00)	0.09 (0.24)	†
B	0.40 (0.37)	0.50 (0.30)	ns
C	0.64 (0.38)	0.73 (0.33)	ns
D	0.93 (0.16)	0.77 (0.31)	*
E	0.23 (0.27)	0.35 (0.32)	ns
F	0.30 (0.45)	0.16 (0.23)	ns
Trick	1.00 (0.00)	0.82 (0.39)	*

Table 4 Results of the situational awareness questionnaire for the San Francisco driving video

Type	Novice drivers M(SD)	Expert drivers M(SD)	F
A	0.55 (0.19)	0.65 (0.18)	†
B	0.38 (0.45)	0.30 (0.42)	ns
C	0.56 (0.36)	0.70 (0.39)	ns
D	0.62 (0.35)	0.84 (0.35)	*
E	0.02 (0.10)	0.09 (0.19)	ns
F	0.64 (0.36)	0.39 (0.45)	*
Trick	0.84 (0.37)	0.73 (0.45)	†

a corner, the mean rate of correct answers was significantly higher for novice than for expert drivers, F (1, 45) = 4.93, $p < 0.05$. The number of participants who answered "yes" to the trick question was significantly higher among expert than among novice drivers: F (1, 45) = 5.32, $p < 0.05$.

Although the main effect of driving experience did not have a significant effect on the situational awareness questionnaire for the San Francisco driving video, a main effect of question type was seen: F (6, 270) = 25.5, $p < 0.01$. A significant interaction between driving experience and type of question was also apparent: F (6, 270) = 2.75, $p < 0.05$. A marginally significant difference in the mean rate of correct answers for type A questions was found, with expert drivers showing significantly higher scores than novices, F (1, 45) = 2.84, $p < 0.10$. The mean rate of correct answers for type D questions showed a similar difference, with expert drivers having significantly higher scores than novices, F (1, 45) = 4.41, $p < 0.05$. In the type F questions about a focal object, the mean rate of correct answer was higher for novice than for expert drivers.

4 Discussion

In this study, we measured Japanese novice and expert drivers' eye movements while driving in their own country and in a foreign country to investigate whether eye movements, fixation location and fixation duration, were affected by driving experience and environment. Participants also answered questions to support the results of the eye tracking.

4.1 Assessing the Validity of the Eye Tracking Data and Analysis with the Results of the Osaka Driving Video

The differences in total fixation duration within the prescribed areas in the Osaka driving video showed the same tendency as a previous study [4–6]: novice drivers tended to gaze at a guard rail on a straight road and to look from the center to the outside edge of the lane when driving on a curve. In the area of the right lane on a straight road, a guard rail was included, and the results for that area showed that novice drivers gazed longer at it than expert drivers. While turning right, expert drivers focused on the right. While turning left, fixation duration within the area on the left was longer in novice drivers than expert ones. On a straight road, novice drivers also focused on the middle in the same situation, and it seems that the result showed the same tendency as a previous study, which showed novice drivers' fixation duration and fixation area were longer and narrower than those of expert drivers, respectively [4].

Although these two tendencies when driving on a curve were the same as those in the previous study [6], these results did not indicate consistency in the two kinds of curve driving scenarios. While turning right, a pedestrian was walking on the left side, in the area on the left. In the situational awareness questionnaire for the Osaka driving video, type B questions, Q9 and Q10, asked about this pedestrian, and the results indicated there was no statistical difference between novice and expert drivers in the mean rate of correct answers. The results of the questionnaire suggest that the pedestrian was an object to look at and it reduced the difference in total fixation duration within the area on the left. While turning left, a big red sign was included in the area of the center and the right. Total fixation duration in the area of the sign showed that expert drivers tended to gaze longer in the area than novice drivers. However, the mean rate of correct answers for type F questions did not differ between experts and novices. These results showed that expert drivers tended to gaze longer at the sign than novice drivers, although the sign attracted eye fixation regardless of driving experience. It seems that the difference in degree of attention to the sign influenced total fixation duration within the area on the left while turning left.

These results and discussion support the validity of this research. However, this subject requires further investigation because we did not analyze all objects in the driving videos and did not investigate the validity of the questionnaire or the manner of dividing the frames.

4.2 Differences in Fixation Duration Based on Driving Environment

A comparison of total fixation duration on various areas in the San Francisco driving video with that for the same areas in the Osaka one revealed that the areas showing statistically significant differences in total fixation duration were completely different in the two. In the San Francisco video showing a straight road, total fixation times on the upper middle area and on buildings to the left differed significantly; expert drivers gazed at these areas longer than novices did. There was no difference in total fixation duration between these scenes. However, expert drivers' total fixation duration, not differentiated by areas and scenes, was shorter in the San Francisco driving video than in the Osaka one. This result shows that expert drivers are influenced by familiarity with the driving environment more strongly than novices are. A previous study [8] indicated that increasing cognitive workload made fixation duration shorter. Thus, expert drivers probably need to think more while driving abroad than while driving in their own country because of the difficulty of adapting themselves to the new situation by overcoming normal patterns. Differences in total fixation duration on the upper middle part of a scene and on buildings to the left may reflect a tendency in expert drivers' eye movements seen in a previous study [6], which showed that expert drivers moved their eyes vertically more than novice drivers did. However, further investigation about the reasons that differences in fixation duration occurred only in these areas is needed.

5 Conclusions

The results of eye tracking indicated that drivers' familiarity with the driving environment was one of the factors affecting eye movements while driving. Expert drivers' fixation duration was shorter in the San Francisco driving video, a driving situation in a foreign country for these drivers, than in the Osaka driving video, a more typical driving situation. Expert drivers were more strongly influenced by the driving environment. This tendency to change fixation duration was also affected by situations in which cognitive workload was increased. Thus, in developing an automated driving system, we need to consider ways of supporting safer driving based on a driver's prior driving experience and background, as driving opportunities in foreign countries are increasing with increasing globalization. The results of this study support the future direction of development of automated driving systems that attend to drivers' characteristics.

Directions for future work include analyzing all objects in the driving video, investigating the validity of the questionnaire, and exploring alternative ways of dividing the frames. Further investigation asking why differences in fixation duration occurred only in the upper middle area of the scene and in buildings on the left is also required. In this study, Japanese subjects participated in the experiment. Assessing cultural differences by studying subjects from other countries will also be included in future work.

Acknowledgements We thank the researchers at the Center for Design Research at Stanford University for helping to film the driving videos in San Francisco.

References

1. NHTSA: National Motor Vehicle Crash Causation Survey: Report to Congress (2008)
2. Erwin, H.: Driver vision requirements. Soc. Automot. Eng. Techn. Pap. Ser. **700392**, 629–630 (1970)
3. SAE On-Road Automated Vehicle Standards Committee. Taxonomy and definitions for terms related to on-road motor vehicle automated driving systems; Technical report J3016_201401. SAE, Hong Kong, China (2014)
4. Mourant, R.R., Rockwell, T.H.: Strategies of visual search by novice and experienced drivers. Hum. Factors: J. Hum. Factors Ergon. Soc. **14–4**, 325–335 (1972)
5. Kimiharu, S.: Visual search and peripheral vision performance by novice and experienced drivers. IATSS Rev. 19–3 (1993)
6. Laya, O.: Eye movements in actual and simulated curve negotiation task. ITASS Res. **16**(1), 15–26 (1992)
7. Miura, T.: Visual attention and the safety in driving. ITE Trans. Media Technol. Appl. **61**(12), 1689–1692 (2007)
8. Miura, T.: Active function of eye movement and useful field of view in a realistic setting. In: Gorner, R., d'Ydewalle, G., Parham, R. (eds.) From Eye to Mind: Information Acquisition in Perception, Search and Reading. Amsterdam, pp. 119–127 (1990)
9. Paeglis, R., Bluss, K., Atvars, A.: Driving experience and special skills reflected in eye movements. Proc. SPIE **8155**, 815516-1 (2011)
10. Olsson, P.: Real-time and Offline Filters for Eye Tracking. KTH Royal Institute of Technology (2007)

On Password Strength: A Survey and Analysis

Gongzhu Hu

Abstract Password has been a predominating approach for user authentication to gain access to restricted resources. The main issue with password is its quality or strength, i.e. how easy (or how hard) it can be "guessed" by a third person who wants to access the resource that you have access to by pretending being you. In this paper, we review various metrics of password quality, including one we proposed, and compare their strengths and weaknesses as well as the relationships between these metrics. We also conducted experiments to crack a set of passwords with different levels of quality. The experiments indicate a close positive correlation between the difficulty of guessing and the quality of the passwords. A clustering analysis was performed on the set of passwords with their quality measures as variables to show the password quality groups.

Keywords Password quality · Entropy · Levenshtein distance · Hashing · Password cracking

1 Introduction

Online security has been a major concern since the time when the Internet became a necessity for the society, from business activities to everyday life of ordinary people. A fundamental aspect of online security is to protect data from unauthorized access. The most commonly used method for doing this is to use password as part of the online access process. Password is a secret character string only the user knows and its hashed code is stored on the server that provides access to the data. When the user requests for data access, he enters the password along with other identification information, such as user name or email. A message digest cypher (hash) of the password is computed and the hashed code is transmitted to the server that matches the hash

G. Hu (✉)
Department of Computer Science, Central Michigan University, Mount Pleasant, MI 48859, USA
e-mail: hu1g@cmich.edu

R. Lee (ed.), *Software Engineering, Artificial Intelligence, Networking and Parallel/Distributed Computing*, Studies in Computational Intelligence 721,
DOI 10.1007/978-3-319-62048-0_12

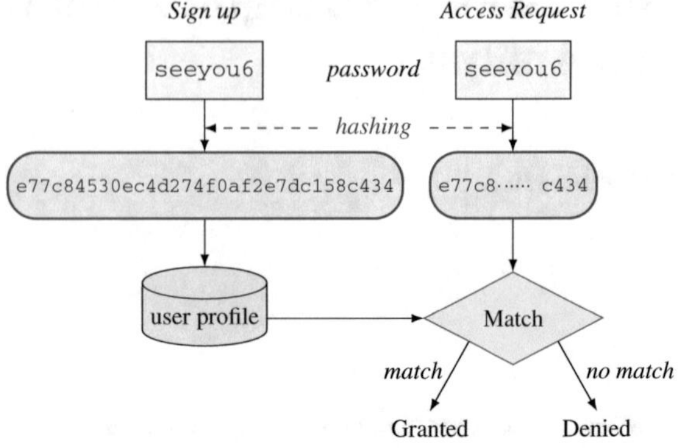

Fig. 1 Password mechanism for resource access request

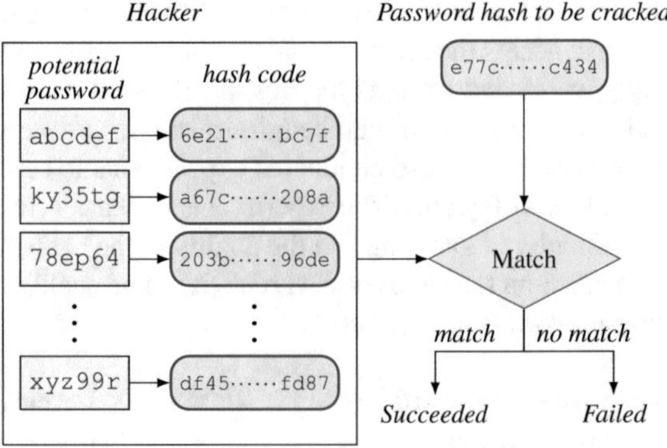

Fig. 2 Hacking the hash code of a password

code previously stored in its database. If a match is found, the user is assumed to be the one who claims who he is, and access is granted; otherwise access is denied, as illustrated in Fig. 1.

Formally, a hash function f maps a password (character string) p to a hash code (cypher) h:

$$f(p) = h \tag{1}$$

where f is irreversible, i.e. function f^{-1} does not exist.

Since the hashing function is irreversible, the only way a third person (a hacker) can guess the password is to try a possible character string to see if its hash code matches the hash code of the actual password. If no match, the hacker may try another string, and continues this process (for off-line attack) until a match is found (the password is cracked) or the hacker gives up without success. The idea of cracking a password is shown in Fig. 2, where the hacker tries a sequence of potential password $p_1, p_2, \ldots, p_k$ to generate hash codes $h_1, h_2, \ldots, h_k$, and to see if $h_k = h$, where h is the given hash code to be cracked. For online attack, a "three strikes" type of rule may prevent the hacker to try more than 3 times, though.

The problem for the hacker to figure out is how to select strings $p_1, p_2, \ldots, p_k$ as potential passwords. There are various approaches for doing so, all based on guessing. From the user's point of view, the critical point to the secrecy of the password is make it very hard for hackers to guess the password from the hash code. This is the issue of *password quality* (or strength).

In this paper, we survey various measures for password quality including a new *complexity* measure that we propose in the paper, and compare these measures by analyzing their relationships. Experiments on cracking a small set of passwords of different quality levels was conducted to confirm the correlation between the difficulty of cracking and password quality. We also applied clustering algorithm on these measures to a set of passwords to group them into four quality clusters (weak, fair, medium, strong). Since password quality is mostly an indicator as to how hard a password can be cracked, we first briefly discuss approaches to crack passwords.

2 Password Cracking Methods

Each password p is hashed using an encryption function f to generate its hash code h as defined in Eq. (1). Since only h (rather than p) is stored in the database on the resource server, the task of cracking a password is to use a method c such that $c(h) = p$. Note that c is not f^{-1} that doesn't exist. So, the hacker needs to figure out what method c to use such that he has a better chance to succeed finding p.

Commonly used methods for cracking passwords include brute-force attack, dictionary attack, and some variations of these with time-space tradeoff considerations.

2.1 *Brute-Force Attack*

Let $p = a_1 a_2 \ldots a_l, a_i \in \Sigma$ be a password of length l, where Σ is an alphabet of character set and $N = |\Sigma|$. Given the hash code $h = f(p)$, brute-force attack is to try each possible string $s = x_1 x_2 \ldots x_l, x_i \in \Sigma$ until $f(s) = h$.

The amount of time t to crack a password is proportional to the number of possible strings:

$$t \in \Theta(N^l) \tag{2}$$

where N^l is the size of string pool the possible passwords are drawn from the alphabet Σ. That is, the time complexity of brute-force attack is *polynomial* of N and *exponential* of l. In general, the alphabet Σ is one of the basic sets or combination of them. The basic sets as given in Table 1.

Most passwords are from lowercase letters (***L***) only. In this case, for a password of length 6 (minimum length required by most service sites), there are $26^6 \approx 308.9 \times 10^6$ possible strings to try. If we can make 100,000 calls to the crypt function f per second, it will take 3089 s, or 51.5 min, to exhaust all possible strings from ***L***. For a larger alphabet, say $\boldsymbol{D} \cup \boldsymbol{L}$, a 6-char password will need over 6 h to crack, while a 7-char password needs over 9 days, and 11 months for a 8-char password.

It is practically infeasible to use brute-force approach to crack a password of length 7 or longer with a larger Σ. It is the basic reason why most service sites require passwords to be at least 8 characters long, with at least an uppercase letter, and a digit or a special character, in addition to lowercase letters. This will make $N = (62 \text{ or } 94)$ and $l \geq 8$.

2.2 Dictionary Attack

Since most people use human-memorable passwords that are likely words in dictionaries or some variations of these words, a hacker can try each word in the dictionary rather than random string used in brute-force. With this approach, each word $w_i \in D$ in a dictionary D is checked to see if $f(w_i) = h$, where h is the given hash code to be cracked. Hence, it is very easy to crack if the password is a word in the dictionary. In practice, all the hash codes $f(w_i)$ are pre-computed and stored in a database, rather than computed at run time.

There are two issues with this approach. First, the dictionary needs to contain a very large number of words to cover most (if not all) passwords that you think people may use. Second, most users are aware of dictionary attack and avoid using actual dictionary words for passwords; instead, they make small changes to the word so that it is still easy to remember. For example, rather than directly using `essay` as password, they may use `essay1`, `e55ay`, or `3ssay`. Hence, dictionary attack often apply some rules to the "spelling" of the words, such as replacing `l` by `1`, `o` by `0`, `e` by `3`, `s` by `5`, etc.

Let D be the dictionary used and m be the number of rules, the time t to crack a password using dictionary attack is

$$t \in \Theta(|D| + m) \tag{3}$$

That is, the time complexity is *linear* to the size of the dictionary and number of rules, a significant improvement from brute-force. However, the method may fail to find the password if the dictionary does not contain the password or its variations after applying the rules.

Table 1 Basic alphabet

<table>
<tr><th>Σ</th><th>N</th></tr>
<tr><td>D = {0..9}</td><td>10</td></tr>
<tr><td>L = {a..z}</td><td>26</td></tr>
<tr><td>U = {A..Z}</td><td>26</td></tr>
<tr><td>S = {~‘!@#$%^&*()-_=+[{]}\|;:”’,<.>/?}</td><td>32</td></tr>
</table>

2.3 Table Lookup and Rainbow Table Attack

To speed up the cracking time, table lookup attack stores precomputed hashes of potential passwords (dictionaries) in a database, and cracks a given password hash by searching the database. The search is a lot faster than the original dictionary attack simply because no need to compute the hash for each guess at run time. However, the method requires a huge amount of storage space to store "all" possible passwords and their hashes.

Rainbow table attack [16] is a variation of table lookup attack. Instead of pre-computing the hash codes of a large number of potential passwords and storing in a database, rainbow table approach is a time-memory trade-off to store much less number of hash codes that still represent a huge number of passwords. The basic idea is to create a password-hash chain of length k that covers k potential passwords and their hash codes.

For each word $w \in D$ in a dictionary D, we create a chain $c = (p_1, h_1, \dots, p_k, h_k)$, where $p_1 = w, h_i = f(p_i)$ and $p_i = r(h_{i-1}), f$ is a crypt hash function and r is a *reduce* function that "reduces" a has code to a potential password. For each chain, only the pair (p_1, h_k), i.e. the *initial* password and the *ending* hash code, is stored in the database. Hence, if $k = 10,000$, the pair (p_1, h_n) represents all the 10,000 passwords and their hash codes in the chain, hence a significant saving on storage space. A rainbow table T is a set of chains: $T = \{c_i, i = 1, \dots, n\}$, where c_i is the chain for word w_i in D.

To crack a given hash code h of password p, we check if h equals the ending hash h_k. If it does, $r(h_k)$ is the target password p. Otherwise, we keep applying r and f alternatively backward in the chain until the password is found, or we move the next chain in the table T. If all chains are exhausted, we just failed to find p.

The time complexity of rainbow table attack is

$$t \in \Theta(k|D|) \tag{4}$$

that is *linear* to the dictionary size but with a constant coefficient k that may be quite large (say, 10,000 or larger). It is a time-memory trade-off that, using the same storage space, it covers k times more words using the same dictionary but about k times slower than dictionary attack.

One of the problems with rainbow table approach is collision when two different hash codes are reduced to the same password. Another difficult is how to make the reduce function r "well behaved." That is, r should map the hash codes into well-distributed (likely as user-selected rather than random) password set.

2.4 Attack Using Markov Model

It was argued that users prefer passwords that are easy to remember. Most users are aware of dictionary attack, so human-memorable passwords are mostly not in dictionaries, nor random (randomly generated passwords are hard to remember). One of the approaches to attack human-memorable passwords is "smart dictionary" attack using dictionaries that contain passwords that users are likely to generate. Narayanan and Shmatikov introduced a fast dictionary attack method based on the likelihood of the sequence of characters in users' passwords [15]. The method uses standard Markov model to generate smart dictionary that is much smaller than the ones used in traditional dictionary attack. The main observation given in [15] is that "the distribution of letters in easy-to-remember passwords is likely to be similar to the distribution of letters in the users' native language." Hence, the Markovian dictionary can be created based on the probability of the characters in a sequence.

Let $\nu(x)$ be the frequency of character x in English text, and $\nu(x_{i+1}|x_i)$ be the frequency of x_{i+1} given that the previously generated character is x_i. In the zero-order Markov model, the probability of a sequence $\alpha = x_1x_2 \ldots x_n$ is

$$p(\alpha) = \prod_{x\in\alpha} \nu(x)$$

where the distribution of each character is independent of the previous character. In the first-order Markov model

$$p(x_1x_2 \ldots x_n) = \nu(x_1) \prod_{i=1}^{n} \nu(x_{i+1}|x_i)$$

Markovian dictionaries are created accordingly at two levels. The zero-order dictionary is defined as

$$D_{\nu,\theta} = \{\alpha : \prod_{x\in\alpha} \nu(x) \geq \theta\}$$

where *theta* is a threshold. The first-order dictionary is

$$D_{\nu,\theta} = \{x_1x_2 \ldots x_n : \nu(x_1) \prod_{i=1}^{n} \nu(x_{i+1}|x_i) \geq \theta\}$$

The great advantage of these models is that they drastically reduce the size of search space be eliminating the majority of words from the dictionary that are not likely to be user-selected passwords. It is shown in [15] that if $\theta = 1/7$ (i.e. only 14% of sequences are produced while 86% of sequences are ignored) the zero-order dictionary still has 90% probability covering the plausible passwords. A dictionary containing 1/11 of the keyspace has 80% coverage and 1/40 of keyspace has 50% coverage. Their experiments using the dictionaries of small fraction of the search space successfully recovered 67.6% of the passwords that is a lot higher than many previous work.

2.5 Attack with Probabilistic Context-Free Grammar

Dictionary attack often uses word-mangling rules, but it can be difficult to chose effective mangling rules. One approach to address this problem is to generate guesses in the order of their probability to be user passwords. This would increase the likelihood of cracking the target password in a limited number of guesses. The basic idea of this approach is to estimate the probability of the user passwords from a training set, a set of disclosed real passwords, and create a context-free grammar to be used to estimate the likelihood of the formation of a string [21, 23].

A probabilistic context-free grammar is defined as $G = (\boldsymbol{V}, \boldsymbol{\Sigma}, T, \boldsymbol{P})$, where $\boldsymbol{V}$ is a finite set of non-terminals (variables), $\boldsymbol{\Sigma}$ is a finite set of terminals, $T \in \boldsymbol{V}$ is the start variable, and $\boldsymbol{P}$ is a set of production rules of the form

$$\alpha \to \beta, (p)$$

where $\alpha \in \boldsymbol{V}$ is a variable, $\beta \in \boldsymbol{V} \times \boldsymbol{\Sigma}$ is a string of symbols (variables and terminals), and p is the probability associated with the production rule in such a way that $\sum_i p_i = 1$ for all productions i that have the same α.

In the case for passwords, the only variables (in addition to the start symbol T) are $\boldsymbol{L}$, $\boldsymbol{D}$, ans $\boldsymbol{S}$, representing letters, digits, and special characters. Notations $\boldsymbol{L}_k$, $\boldsymbol{D}_k$ and $\boldsymbol{S}_k$ represent consecutive k letters, consecutive k digits, consecutive k special symbols, respectively. The probability p_i of each production rule i is estimated using the training set. The probability of a sentential form (a string derived from T) is the product of the probabilities of the productions used in the derivation. An example of a derivation is

$$S \Rightarrow \boldsymbol{L}_3\boldsymbol{D}_1\boldsymbol{S}_1 \Rightarrow \boldsymbol{L}_3 4\boldsymbol{S}_1 \Rightarrow \boldsymbol{L}_3 4\#$$

in which production rules $\boldsymbol{D}_1 \to 4$ and $\boldsymbol{S}_1 \to \#$ are used.

The pre-terminal structures (sentential forms) are ordered in decreasing probability, and dictionary words and hashes can be filled in for guessing.

3 Password Quality Measures

Analysis of password strength has been an active area for research and practice for a long time. The focus of these work is on the metrics of password strength and evaluation of these metrics. We shall survey several metrics for password quality including *complexity* that we propose here.

3.1 Entropy

Entropy is a measure of uncertainty and was a term used by Claude Shannon in his information theory [19]. He applied entropy to analysis of English text as "the average number of binary digits required per letter of the original language" [20]. The entropy H of variable X is defined as

$$H(X) = -\sum_{x} p(X = x)\log_2(X = x)$$

where $p(X = x)$ is the probability of X's value being x.

National Institute of Standards and Technology (NIST) published *Electronic Authentication Guidelines* [3] that provided a metric for estimating password entropy as a measure of password strength. In the NIST Guidelines, entropy denotes the uncertainty of a password, expressed as number of bits. The high the entropy value, the higher uncertainty of the password, meaning that the password would be harder to guess. For a randomly selected password w of length m from a charset of size N, the entropy is defined in Eq. (5) below.

$$H(w) = \log_2(N^m) = m \times \log_2 N \tag{5}$$

It is obvious that longer passwords from larger charsets will have higher entropy values.

The NIST Guidelines gave an outline to estimate the entropy of user selected password based on the length of password, charsets, and possibility of dictionary attack. It mostly assigns additional bits to the entropy as the password's length increases, and added certain "bonus" bits to the use of multiple charsets as well as dictionary check. The estimated entropy H is given below.

$$H = \begin{cases} +4 & \text{first character} \\ +2 & \text{for each of the 2nd to 8th characters} \\ +1.5 & \text{for each of the 9th to 20th characters} \\ +6 & \text{use uppercase and non-alphabetic characters} \\ +6 & \text{not in dictionaries} \end{cases}$$

For the last two bonus bits, there may involve other composition rules and properties of dictionary check.

3.2 Password Quality Indicator

Most users are aware of dictionary attack and avoid using dictionary words for passwords. However, users want passwords easy to remember, so they tend to use a word and make small changes to it. With this consideration, methods of dictionary attack also adopted various word-mangling rules to match a password with words in dictionaries. So, the strength of a password not only considers if the password is in dictionaries, but also should measure how easy (or hard) to correct the "spelling errors" in the password so it can match some words in the dictionaries. This is a commonly measure as a linguistic distance between the password and a dictionary word. The very basic linguistic distance is the *Levenshtein distance* (or edit distance) that is the minimum number of editing operations (insert, delete, and replace) needed to transfer one word to another. This idea was used in the *password quality index* (PQI) metric proposed in [13] and refined in [14].

The PQI of a password w is a pair $\lambda = (D, L)$, where D is the Levenshtein distance of w to the base dictionary words, and L is the effective length of p, which is defined as

$$L = m \times \log_{10} N \tag{6}$$

where m is the length of w and N is the size of the charset where the characters of w is drawn from.

The effective length is the length calculated in a "standardized" charset, the digit set $\boldsymbol{D}$ given in Table 1. The idea behind this is that a password (e.g. `k38P` of length 4) drawn from multiple charsets ($\boldsymbol{D} \cup \boldsymbol{L} \cup \boldsymbol{U}$) is just as hard to crack (or, has about the same number of possible candidates to crack) as another password (e.g. `378902` of length 6) from only the digit set $\boldsymbol{D}$.

It is seen that the effective length of a password given in (6) is essentially the same as the entropy value in (5), just off by a constant factor $\log_2 10$. Both consider the password's length as well as the size of the charset.

With the PQI measure, The quality criterion given in [13] states that a password is of good quality if $D \geq 3$ and $L \geq 14$.

3.3 Password Strength Meters by Service Vendors

Service vendors use various meters to measure password's strength with somewhat different algorithms. A good survey and analysis of the meters were given in [4] that listed the basic requirements at these vendors, partly shown in Table 2.

Some of these vendors also use user information (such as surname) in their classification of password quality. Most of these meters are based on the traditional LUDS requirements (**u**ppercase and **l**owercase letters, **d**igits, and **s**pecial chararacters), except Dropbox.

Table 2 Password requirements at various vendors [4]

Service	Strength scale	Length limits		Charset required
		Min	Max	
Dropbox	Very weak, Weak, So-so, Good, Great	6	72	∅
Drupal	Weak, Fair, Good, Strong	6	128	∅
FedEx	Very weak, Weak, Medium, Strong, Very strong	8	35	1+ lower, 1+ upper, 1+ digit
Microsoft	Weak, Medium, Strong, Best	1	–	∅
Twitter	Invalid/Too short, Obvious, Not secure enough could be more secure, Okay, Perfect	6	> 1000	∅
Yahoo!	Weak, Strong, Very strong	6	32	∅
eBay	Invalid, Weak, Medium, Strong	6	20	any 2 charsets
Google	Weak, Fair, Good, Strong	8	100	∅
Skype	Por, Medium, Good	6	20	2 charsets or upper only
Apple	Weak, Moderate, Strong	8	32	1+ lower, 1+ upper, 1+ digit
PayPal	Weak, Fair, Strong	8	20	any 2 charsets[a]

[a]PayPal counts uppercase and lowercase letters as a single charset

Dropbox's password strength checker, called *zxcvbn* [24], uses a different approach to estimate the strength of a password. The basic idea is to check "how common a password is according to several sources." The sources include common passwords in leaked password sets, common names from census data, and common words in Wikipedia. The *zxcvbn* algorithm finds patterns (sub-strings) in a password that match items in the sources, and these patterns may overlap within the password. The patterns include token (`logitech`), reversed (`DrowssaP`), sequence (`jklm`), repeat (`ababab`), keyboard (`qAzxcde3`), date (`781947`), etc. It then assigns a guess attempt estimation to each match, and finally searches for non-overlapping adjacent matches that cover the password and has the minimum total guess attempt.

Table 3 Password multi-checker output for `password$1` [4]

Service	Strngth score	
Apple	Moderate	2/3
Dropbox	Very weak	1/5
Drupal	Strong	4/4
eBay	Medium	4/5
FedEx	Very weak	1/5
Google	Fair	3/5
Microsoft (v3)	Medium	2/4
PayPal	Weak	2/4
Skype	Poor	1/3
Twitter	Perfect	6/6
Yahoo!	Very strong	4/4

The algorithms used by these vendors resulted in very diverse strength scores for the same passwords. For example, as illustrated in [4], the password `password$1` scored from very weak to very strong, given in Table 3.

3.4 Password Complexity Metric

We propose a password *complexity* metric that considers both the common LUDS requirements and the patterns in the password. As in many other measures, the number of different charsets used in a password still plays an important role in this metric. In addition, we consider other factors that may make a password harder to guess, such as mix of same-charset-substrings, position of special symbols, and substrings in the dictionary.

As the users are aware of using different charsets to compose passwords, they are likely to put characters from the same charset together rather than mixing the them up. For example, it is more likely to have a password `horse743` instead of `ho7r4se3`. The latter is considered more "complex" than the former and harder to crack. We find substrings in a password that are from the same charset, and count the number of such substrings. Using the same example, the number of substrings in `horse743` is 2 (`horse` and `743`), while the number of substrings in `ho7r4se3` is 6 (`ho`, `7`, `r`, `4`, `se`, and `3`). However, this number may be higher for a longer password than a shorter one, so we take the ratio of this number vs the password's length as a factor to our metric.

Another factor is the position of special symbols. Many users use a common word and then add a special symbol at the end (or beginning). For example, `horse#` may be more common than `hor#se`. We apply a small penalty to this pattern if the password uses only 2 charsets (including the special symbol charset).

A password that matches a dictionary word is weak, but a password with a substring that matches a dictionary word may or may not be weak, depending on the length of the substring itself and its "weight" in the password. Many user-selected passwords contain substrings that are dictionary words but the passwords may be good. For example, `Pfan?6tk` is pretty strong by most of the measures we discussed, although it contains a dictionary word `fan`. However, a longer dictionary word (4 letters or more) in a password would make it weaker, particularly for a relatively shorter password. The password `Wfoot67` is a lot weaker that `Wdfoot6237`, with both containing the dictionary word `foot`.

With these considerations, our complexity metric of password w is calculated as given in Eq. (7).

$$C(w) = \begin{cases} 0 & \text{if } l < 4 \\ n + (k/l) + s - p - (d/l) & \text{if } l \geq 4 \end{cases} \tag{7}$$

where n is the number of charsets in w, k is the number of same-charset-substrings, l is the length of w, s is a bonus if w has special characterize, p is special character position penalty, and d/l is in-dictionary penalty. In particular, $s = 0.5$ if w contains special character, 0 otherwise; $p = 0.5$ if a special character is at the beginning or the end of w and w has no more than 2 charsets, 0 otherwise; d is the length of substring that is a dictionary word.

We may scale the metric up from the range of 0–10 to 0–100 (by a factor of 10) so it can be reasonably compared with other measures such as NIST entropy.

4 Experiments and Analysis

We conducted some preliminary experiments to compare these password strength measures. Several datasets were used in our experiments. First, we calculated some statistics of the relationships between the measures. Then, we applied clustering algorithm to passwords in the r-dimensional metric space where r is the number of metrics to see if the passwords are well grouped in strength-clusters. Finally, we tried to crack some of the passwords to confirm the relationship between the strength and the level of difficulty to crack them.

4.1 Data Sets

Three data sets were used in our experiments as explained in Table 4. The data sets D_1 and D_2 were primarily for correlation analysis, whereas D_3 was used to practice password cracking to illustrate the strength of the passwords.

4.2 Statistical Analysis of Strength Metrics

Two statistics, *Pearson correlation* and *maximum information coefficient* (*MIC*) were calculated to show the relationships between the strength measure. MIC was introduced in [18] as a new exploratory data analysis indicator that measures the strength of relationship between two variables. This measure is a statistic score between 0 and 1. It captures a wide range of relationships and is not limited to specific function types (such as linear as Pearson correlation does).

We calculated the Pearson correlation and MIC on four measures (entropy, NIST entropy, Levenshtein distance, and complexity). The pairwise results for data set D_1 and D_2 are given in Table 5(a) and (b), respectively.

It shows that the entropy, NIST entropy, and complexity have high MIC scores, indicating that the three measures are closely related and not very independent. Their

Table 4 Data sets used in experiments

Data	Size	Description of passwords in the set
D_1	10,000	Most commonly used [9]
D_2	642	Mixed user-elected and randomly generated
D_3	127	Manually selected with various composition patterns

Table 5 MIC and Pearson coefficients between measures

(a) for data set D_1

Measure 1	Measure 2	MIC	Pearson Correlation
Entropy	NIST entropy	0.79863	0.72584134
Entropy	Complexity	0.84558	0.08722651
NIST entropy	Complexity	0.89118	0.40052018
Levenshtein	Entropy	0.26666	−0.00422713
Levenshtein	NIST entropy	0.33358	0.52798074
Levenshtein	Complexity	0.48328	0.56144760

(b) for data set D_2

Measure 1	Measure 2	MIC	Pearson correlation
Entropy	NIST entropy	0.93002	0.88889660
Entropy	Complexity	0.90081	0.80279726
NIST entropy	Complexity	0.89208	0.90321730
Levenshtein	Entropy	0.70443	0.74756646
Levenshtein	NIST entropy	0.69875	0.84680110
Levenshtein	Complexity	0.66447	0.78521204

MIC scores with Levenshtein measure, though, are a lot lower indicating that it is relatively independent to the other three measures. This may confirm the claim in PQI [13] that Levenshtein distance of the password and words in the dictionaries may indeed be an alternative (and independent) to the LUDS requirements. It is interesting to note that the linear correlation between entropy and complexity is close to 0, so is the correlation between entropy and Levenshtein distance.

The p-value of the MIC statistic for the data set of size $n \geq 600$ with $MIC \geq 0.2257$ is very small ($p = 0.00001281 \pm 10^{-6}$ with $\alpha = 0.05$), indicating that the correlation of any pair of measures is significant. Here 0.2257 is the largest MIC value in the p-value table that is smaller than 0.26666, the smallest in Table 5(a).

On data set D_2, the linear correlation and the MIC scores are much higher. This may be due to composition of the patterns in the passwords in this data set.

From the correlation analysis, we see that most of these measures (entropy, NIST entropy, complexity) are highly correlated. This may lead to the conclusion that the password strength classifications using any of these measures will not differ much. The Levenshtein distance of a password from dictionary words, however, is a different measure (less closely correlated to the other measures), that may be used for estimating password's strength along with other measures, ad did in PQI [13].

4.3 *Password Strength Test*

To further evaluate the password strength metrics, we experimented cracking the passwords in the small data set D_3 with 127 passwords that are manually selected to include strengths as various levels.

4.3.1 Lookup Table and Rainbow Table Attack

We used two online password cracking sites to uncover the passwords (given their MD5 hashes) in the data set D_3. One site is CrackStation that "uses massive pre-computed lookup tables to crack password hashes" [5]. The other is HashKiller [11] that also uses very large hash databases to crack password hashes. The sizes of the lookup table databases of the two sites are listed in Table 6(a), and the results are given in Table 6(b).

All the 17 MD5 hashes that HashKiller failed also failed by CrackStation. HashKiller was more successful than CrackStation simply because it has larger hash databases. These are hashes of passwords that were randomly generated (like `nXXdHtt6Q`, `2y!3e)!%)`, `fwtC9xcO`, etc.). The average strength metrics of the passwords are given in Table 6(c). It is clearly shown that the passwords the sites failed to crack are those with higher strength than those cracked passwords on all the measures. It is also seen that HashKiller was able to crack passwords with higher strength than CrackStation.

4.3.2 John the Ripper Attack

John the Ripper (JtR) [17] is an open source software for cracking passwords. It combines dictionary attack, rainbow table attach, and brute force attack, with various rules for word composition. The free version of JtR comes with a very small dictionary (password list) containing only 3,546 words. We used JtR to crack some passwords in the data set D_3. The time spent on some of these passwords is given in Table 7. The last two passwords in the table took too long to crack and we aborted JtR before the passwords were recovered.

Figure 3 shows plots of various strength measures vs time spent by JtR to crack the passwords.

Since the default dictionary of JtR is very small and we did not use an additional dictionary, only 6 passwords (or slightly changed version of the passwords by some transformation rules in JtR) found in the dictionary very quickly. The other passwords were cracked using the brute-force approach and consumed from 10 min to 1.5 h for each password. From Fig. 3, we can see that the strength metrics entropy and complexity have positive correlations with the time spent by JtR, and the Levenshtein distance measure is even more closely correlated to the time needed to crack the passwords.

Table 6 Crack passwrods on CrackStation and HashKiller

(a) Database sizes

Hash	CrackStation	HashKiller
MD5	190 GB, 15-billion entries	829.7 GB
SHA1		312.0 GB
SHA256	19 GB, 1.5-billion entries	
NYLM		312.0 GB

(b) Cracking results

# MD5 Hashes	CrackStation		HashKiller	
	cracked	%	cracked	%
127	58	45.7%	110	86.6%

(c) Strength metrics of un-cracked and cracked passwords

Measure	CrackStation		HashKiller	
	un-cracked	cracked	un-cracked	cracked
entropy	35.75	23.73	40.02	28.76
NIST entropy	27.84	19.11	31.56	22.66
complexity	38.33	13.53	42.92	24.54
Levenshtein	4.62	1.53	5.94	2.79

Table 7 Time spent on cracking some passwords in D_3 using JtR

Id	Password	MD5	Entropy	NIST	Complexity	Levenshtein	JtR (h:m:s)
1	kitty	`cd880b726e0a0dbd4237f10d15da46f4`	16.29	12.00	4.00	0.00	0:00:00
2	susan	`ac575e3eecf0fa410518c2d3a2e7209f`	16.29	18.00	14.00	1.00	0:00:00
3	jellyfish	`9787922f286a4349a009237d0b2ffd73`	29.32	19.50	2.22	0.00	0:16:12
4	smellycat	`5a7c5156c0f64867b607abf66f3bbe1f`	29.32	22.17	6.67	3.00	1:33:20
5	allblacks	`6186e17cdef7a72b6b44c374b7ade779`	29.32	22.83	7.78	4.00	0:02:56
6	usher	`0d5e410c96a560d494fe2b3485f5f864`	16.29	12.00	4.00	0.00	0:01:54
108	seeyou6	`e77c84530ec4d274f0af2e7dc158c434`	25.08	19.43	21.43	3.00	0:10:15
109	funlo_	`9fd0d44ade138efcb67a69c68b006f08`	24.36	17.00	26.67	3.00	0:24:15
110	Yt	`b6cc855441c4c688509ab84a34722868`	7.90	12.00	0.00	1.00	0:00:18
111	Password1	`2ac9cb7dc02b3c0083eb70898e549b63`	37.14	28.17	31.11	1.00	0:00:19
119	billabong	`c61a71b89ce304b78bce33cb56f9cffa`	29.32	22.83	7.78	3.00	0:00:00
120	Basketball	`c71a18083d9e74b4a5c5d8d9a17d68d0`	39.51	24.60	20.00	0.00	0:00:00
121	phoenix09	`6aed5d571daa97833033534a4ac2c27a`	32.25	20.83	16.67	2.00	0:05:17
126	akjuwfg	`015c09cba7ac3cdfe01e2489e2b8ad14`	22.81	22.00	12.86	4.00	1:07:20
18	oQ4cf5BDA	`3567b3ab8dbbbc1f091e284caaf7976b`	41.27	33.00	42.00	6.00	
127	D$f9	`ecad9ace9534cdb71d04a600960ca3c4`	18.17	22.00	65.00	2.00	

Fig. 3 Time spent by JtR

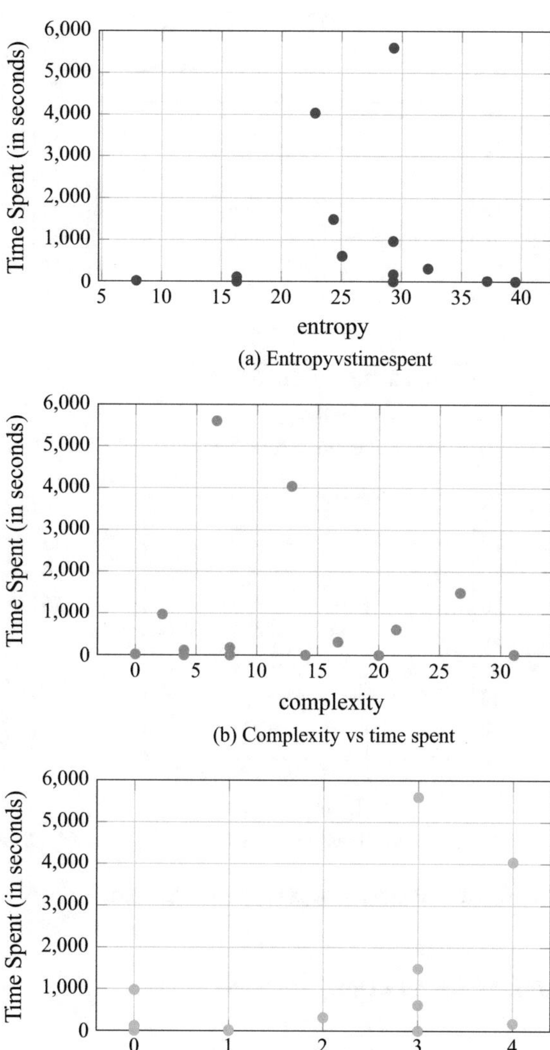

(a) Entropyvstimespent

(b) Complexity vs time spent

(c) Levenshtein distance vs time spent

Table 8 Clutering result

Number of iterations:			24
Within cluster SSE (sum of squared errors):			7.23
Clustered instances	cluster	# instances	%
	1	33	26%
	2	30	24%
	3	42	33%
	4	22	17%

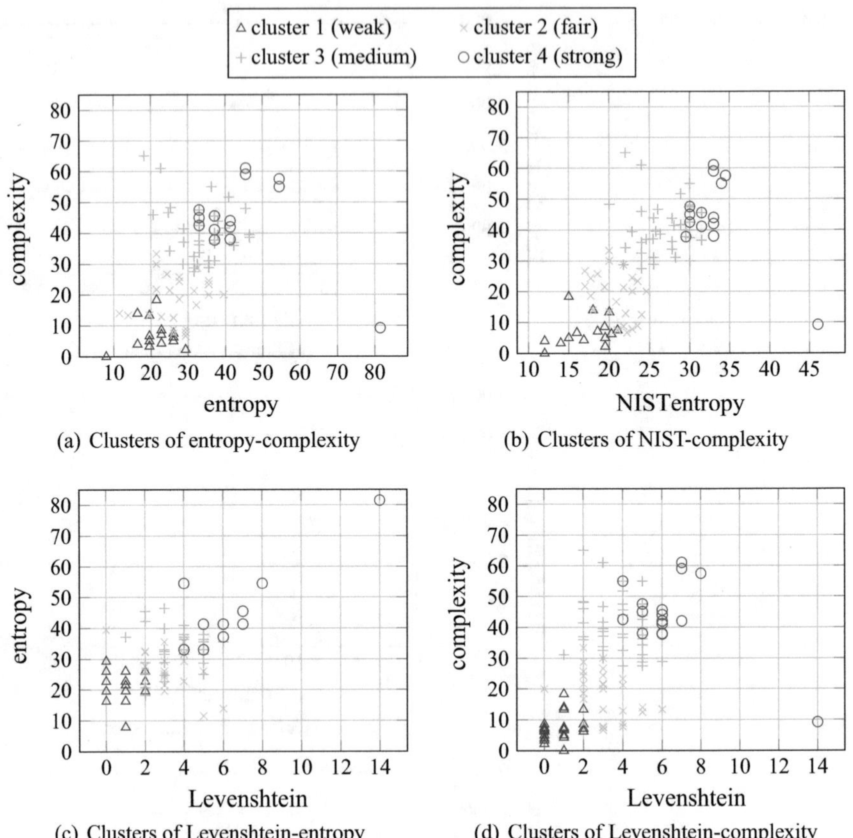

(a) Clusters of entropy-complexity

(b) Clusters of NIST-complexity

(c) Clusters of Levenshtein-entropy

(d) Clusters of Levenshtein-complexity

Fig. 4 Clustering plots on pair-wise strength metrics

4.4 Clustering of Passwords

As another evaluation of the strength metrics is to do clustering of the passwords to see if the weak passwords are grouped together and the strong ones are grouped in another cluster. In addition to the 4 variables (entropy, NIST, complexity, and Levenshtein distance), we also used two more variables when we applied the clustering algorithm. The two variables are *effective length* and a variation of the Levenshtein distance, which calculates the Lavenshtein distance of sub-words in a password against dictionary words, rather than treating the password as a single word.

We applied the K-means clustering algorithm [10] on the 127 passwords in the data set D_3 with the 6 variables into 4 clusters representing four strength levels. The results is summarized in Table 8.

The algorithm randomly selected 4 passwords (with ids 29, 35, 38, and 44) as the initial centroids, ran through 24 iterations and stabilized in 4 clusters with the within-cluster SSE (sum of squared errors) 7.23.

The projections of the 6-D space onto 2-D of some pairs of the metrics are shown in Fig. 4a–d. Note that the number of points of a cluster in the plot may appear smaller that the number in Table 8. For example, there are 22 instances in cluster 4, but only 10–14 cluster-4 points in the plots. This is because of 6-D to 2-D projection and multiple points were mapped to the same spot in the 2-D space. From theses figures, we can visually see that the passwords were reasonably clustered into strength groups, from weak to strong.

5 Related Work

A lot work has been done in assessing the quality of passwords. As mentioned before that a good survey of password meters used by service vendors was given in [4]. Florêncio and Herley studied a set of half-million user passwords (as well as other information about the user accounts) on various web sites to find the characteristics about the passwords and their usages [7]. Kelley et al. analyzed 12,000 passwords and developed methods for calculating the time needed for several password-guessing tools to guess the passwords so that the password-composition policies were better understood [12]. Dell'Amico et al. conducted an empirical analysis of password strength [6], in which the authors analyzed the success rate (number of cracked passwords versus search space) using different tools including brute force, dictionary attack, dictionary mangling, probabilistic context-free grammars [23], and Markov chains [15]. They concluded that the probability of success guessing a password at each attempt decreases roughly exponentially as the size of search space grows. An extension of the probabilistic context-free grammar approach was given in [26] that tries to find new classes of patterns (such as keyboard and multi-word) from the training set. Weir et al. tested the effectiveness of using entropy (NIST) as a measurement of password strength [22]. Their experiments on password cracking techniques on several large data sets (largest had 32 million passwords) showed that the NIST entropy does "not provide a valid metric for measuring the security provided by password creation policies."

Traditional advice to the users to select their passwords does not provide additional security, argued by Florêncio et al. [8]. They pointed out that relative weak passwords are sufficient to prevent brute-force attack on an account if "three strikes" type rule applies. Thus, making password stronger does little to address the real threats. Bonneau used statistical guessing metrics to analyze a large set of 70 million passwords [1]. The metrics is to compare the password distribution with uniform distribution "would provide equivalent security against different forms of guessing attack." The study found little variation in the password distribution that produces "similar skewed distributions with effective security varying by no more than a few bits." Bonneau and his colleagues pointed out the gaps between the academic research on password strength evaluation and the authentication methods used in the reality of the web environment today [2].

The probabilistic context-free grammar method proposed in [23] appears quite effective. A series of experiments based on training sets of relatively large number of disclosed passwords were conducted showing that the approach was able to crack 28–129% more passwords than the traditional John the Ripper method.

Another interesting work about the quality of passwords was given in [25]. The authors conducted psychological and statistic experiments with 288 students regarding their selections of passwords and memorability of these passwords. The students were divided into three groups: *control* group (traditional advice—at least 7 characters long and at least one non-letter), *random* group (random pick 8 characters with eyes closed), and *pass phrase* group (select password based on mnemonic phrases). The experiments lasted for several months (to test memorability) and various methods were used to attack the passwords. The results of their experiments showed that random passwords are hard to remember; passwords based on mnemonic phrases are harder to guess, just as strong an random passwords and as easy to remember as naively selected ones.

6 Conclusion

In this paper we reviewed the basic metrics for assessing password strength, including entropy, NIST entropy, password quality index, and Lavenshtein distance, as well as some password quality metrics developed at some popular service vendors. We proposed a password complexity metric that considers the composition patterns in a password in addition to the LUDS criteria. The correlations between these measures were analyzed using the maximum information coefficient (MIC) and Pearson coefficient. The resulting statistics show that most of the metrics are closely correlated except Levenshtein distance that may be a good measure for password quality besides the traditional parameters (length of password and size of charset).

There password strength metrics were evaluated by experiments that tried to crack the hashes of a small set of passwords to see if the difficulty of cracking a password is indeed related to the strength measures. The cracking tools used including techniques like brute force, transformation rules, dictionary attacks, and massive table look-up. Two types of results were obtained: password found or not (table look-up), and the length of time spent to discover the password (other techniques used in JtR). Results showed that the level of success cracking the passwords is highly positively related to the strength measures. It is pretty convincing from our experiments that the strength metrics are valid and can be used to assess the quality of user selected passwords. This was further validated by the clustering results.

The bottom line of our analysis comes back to the simple advice to users: select a long password with various kinds of characters (lower and uppercase letters, digits, and symbols), as the length of password and size of charset being the two most critical parameters to the strength of the password in all the metrics we studied.

References

1. Bonneau, J.: The science of guessing: analyzing an anonymized corpus of 70 million passwords. In: 2012 IEEE Symposium on Security and Privacy, pp. 538–552. IEEE (2012)
2. Bonneau, J., Herley, C., van Oorschot, P.C., Stajano, F.: Passwords and the evolution of imperfect authentication. Commun. ACM **58**(7), 78–87 (2015)
3. Burr, W.E., Dodson, D.F., Newton, E.M., Perlner, R.A., Polk, W.T., Gupta, S., Nabbus, E.A.: Draft NIST special publication 800–63-2: electronic authentication guideline. US Department of Commerce, National Institute of Standards and Technology (2013)
4. de Carné de Carnavalet, X., Mannan, M.: From very weak to very strong: analyzing password-strength meters. In: Network and Distributed System Security Symposium (NDSS 2014). Internet Society (2014)
5. Defuse Security: Crackstation: Free password hash cracker. https://crackstation.net/
6. Dell'Amico, M., Michiardi, P., Roudier, Y.: Password strength: an empirical analysis. In: 29th IEEE International Conference on Computer Communications, pp. 983–991 (2010)
7. Florencio, D., Herley, C.: A large-scale study of web password habits. In: Proceedings of the 16th International Conference on World Wide Web, pp. 657–666. ACM (2007)
8. Florêncio, D., Herley, C., Coskun, B.: Do strong web passwords accomplish anything? In: 2nd USENIX Workshop on hot Topics in Security (2007)
9. GitHub: 10k most common passwords. https://github.com/danielmiessler/SecLists/blob/master/Passwords/10k_most_common.txt
10. Han, J., Pei, J., Kamber, M.: Data mining: concepts and techniques, 3rd edn. Elsevier (2012)
11. HashC: Hash killer. https://www.hashkiller.co.uk/
12. Kelley, P.G., Komanduri, S., Mazurek, M.L., Shay, R., Vidas, T., Bauer, L., Christin, N., Cranor, L.F., Lopez, J.: Guess again (and again and again): measuring password strength by simulating password-cracking algorithms. In: 2012 IEEE Symposium on Security and Privacy (SP), pp. 523–537. IEEE (2012)
13. Ma, W., Campbell, J., Tran, D., Kleeman, D.: A conceptual framework for assessing password quality. Int. J. Comput. Sci. Netw. Secur. **7**(1), 179–185 (2007)
14. Ma, W., Campbell, J., Tran, D., Kleeman, D.: Password entropy and password quality. In: 4th International Conference on Network and System Security (NSS), pp. 583–587. IEEE (2010)
15. Narayanan, A., Shmatikov, V.: Fast dictionary attacks on passwords using time-space tradeoff. In: Proceedings of the 12th ACM Conference on Computer and Communications Security, pp. 364–372. ACM (2005)
16. Oechslin, P.: Making a faster cryptanalytic time-memory trade-off. In: Annual International Cryptology Conference, pp. 617–630. Springer (2003)
17. Openwell: John the Ripper password cracker. http://www.openwall.com/john/
18. Reshef, D.N., Reshef, Y.A., Finucane, H.K., Grossman, S.R., McVean, G., Turnbaugh, P.J., Lander, E.S., Mitzenmacher, M., Sabeti, P.C.: Detecting novel associations in large data sets. Science **334**(6062), 1518–1524 (2011)
19. Shannon, C.E.: A mathematical theory of communication. Bell Syst. Tech. J. **27**, 379–423 (1948)
20. Shannon, C.E.: Prediction and entropy of printed english. Bell Labs Tech. J. **30**(1), 50–64 (1951)
21. Weir, C.M.: Using probabilistic techniques to aid in password cracking attacks. Ph.D. thesis, Florida State University (2010)
22. Weir, M., Aggarwal, S., Collins, M., Stern, H.: Testing metrics for password creation policies by attacking large sets of revealed passwords. In: Proceedings of the 17th ACM Conference on Computer and Communications Security, pp. 162–175. ACM (2010)
23. Weir, M., Aggarwal, S., De Medeiros, B., Glodek, B.: Password cracking using probabilistic context-free grammars. In: 30th IEEE Symposium on Security and Privacy, pp. 391–405. IEEE (2009)
24. Wheeler, D.L.: zxcvbn: low-budget password strength estimation. In: Proceedings of the 25th USENIX Security Symposium, pp. 157–173 (2016)

25. Yan, J., Blackwell, A., Anderson, R., Grant, A.: Password memorability and security: empirical results. IEEE Secur. Priv. **2**(5), 25–31 (2004)
26. Yazdi, S.H.: Probabilistic context-free grammar based password cracking: attack, defense and applications. Ph.D. thesis, Florida State University (2015)

Probabilistic Model-Based Multistep Crossover Considering Dependency Between Nodes in Tree Optimization

Kohei Matsumura, Yoshiko Hanada and Keiko Ono

Abstract Deterministic Multistep crossover fusion (dMSXF) is one of promising crossover methods of a tree-based genetic programming. dMSXF performs a multi-step local search from a parent in the direction approaching the other parent. In the local search, neighborhood solutions are generated by operators based on a replacement, an insertion and a deletion of nodes to combine both parents' small traits step by step. In our previous work, we improved the search efficiency of dMSXF by introducing a probabilistic model constructed by the search information to the generation of neighborhood solutions. In the method, the probabilistic model considers nodes individually and a node dependency is ignored. The method has a room to further improve if a probabilistic model that can treat information about a dependency relationship between nodes. In this paper, we introduce a probabilistic model which considers the dependency between a parent node and a child node. The search performance of the proposed method is evaluated on symbolic regression problems.

1 Introduction

Genetic programming (GP) is one of promising program evolution algorithms based on evolutionary computation (EC). Similar to the search framework of genetic algorithms (GAs), most GPs perform a selection, a crossover and a mutation to develop a program. Among the reproduction operators, the crossover has been considered to be the key component of evolutionary computations. Especially in optimizing combinatorial structures such as a graph, it is important to design the operator that takes

K. Matsumura
Graduate School of Science and Engineering, Kansai University, Osaka, Japan

Y. Hanada (✉)
Faculty of Engineering Science, Kansai University, Osaka, Japan
e-mail: hanada@kansai-u.ac.jp

K. Ono
Department of Electronics and Informatics, Ryukoku University, Shiga, Japan
e-mail: kono@rins.ryukoku.ac.jp

R. Lee (ed.), *Software Engineering, Artificial Intelligence, Networking and Parallel/Distributed Computing*, Studies in Computational Intelligence 721,
DOI 10.1007/978-3-319-62048-0_13

account of problem-specific structures to keep parents' preferable traits. In optimizing tree structured problems, the design of the crossover operator is an important issue. Various kinds of recombination mechanisms to treat structures of subtrees as a kind of building blocks have been discussed [1–3]. To improve search abilities of GP, dependencies between nodes or semantics of subtrees have been considered, based on symbolic information, in reproduction operators [4–7].

In solving combinatorial optimization problems involving complex constraints by ECs, local searches have been incorporated to crossover in order to adjust the structural details of solutions, since a simple recombination often brings a drastic change in solutions and might break favorable characteristics [8–10]. Deterministic Multistep crossover fusion (dMSXF) is successful one of such a memetic crossover for combinatorial structures [9, 10]. dMSXF performs a sequence of local search which gradually moves the offspring from its initial point to the other parent. In our previous work, we proposed a neighborhood generation based on a replacement, an insertion and a deletion of nodes for tree structures. The neighborhood solutions approach the other parent by getting its traits little by little with three operations in the local search of dMSXF. Due to this mechanism, it can generate a wide variety of solution between parents with keeping their traits [11]. In addition, we incorporated a probabilistic model into the neighborhood generation, in order to improve efficiency of heredity of parents' good traits. Through numerical experiments with several instances of the symbolic regression problem, it was shown that the improved dMSXF finds optimal solutions with high probability compared to conventional dMSXF [12]. However, the independent model used in Probabilistic Incremental Program Evolution (PIPE) that treats a node individually was adopted, and a dependency relationship between node was ignored in the neighborhood generation. In many cases of tree structured problem, the dependence between nodes, especially a parent and a child, is high, which has an intensified impact on the fitness value of a solution.

In this paper, we propose a new neighborhood generation based on the model that considers a dependency relationship between the parent node and the child node that is used in Estimation of Distribution Programming (EDP) [13]. To show the effectiveness of the new neighborhood generation of dMSXF, we evaluate the search performance in symbolic regression problems. In addition, we show that lethal traits that strongly degrade the fitness value of a solution are reduced by considering the parent-child dependency relationship.

2 Deterministic Multistep Crossover Fusion

We consider the problem with minimizing a function $f(x)$. dMSXF performs a local search from parent p_a in the direction approaching the target parent p_b. In the search from p_a to p_b, the k_{max} steps local search centering on the search point x_k ($k = 1, \ldots, k_{max}$) is performed, where the initial point x_1 is p_a. Let $d(\cdot, \cdot)$ denote a distance between two solution. All neighborhood candidates, y, generated from x_k in the local search should satisfy the distance constraint $d(y, p_b) < d(x_k, p_b)$. The distance metric for tree structures is defined in the next section. Let $\mathcal{N}_{\rightarrow p_b}(x_k)$ denote the set of biased neighborhood generated from the search point x_k toward the target solution p_b under the distance constraint. The set of offspring generated by parents p_a and p_b is represented as $C(p_a, p_b)$. The crossover algorithm of dMSXF is as follows:

Procedure deterministic Multistep Crossover Fusion

1 : Set $\mathscr{C}(p_a, p_b) \leftarrow \emptyset$.
2 : Set $k \leftarrow 1$.
3 : Set $\mathscr{C}(p_a, p_b) \leftarrow \mathscr{C}(p_a, p_b) \cup \{p_a\}$.
4 : Set $x_k \leftarrow p_a$.
5 : If $k \leq k_{max}$ and $x_k \neq p_b$, then perform Steps 6–11, otherwise stop.
6 : Generate a set of μ neighborhood candidates $\mathscr{N}_{\rightarrow p_b}(x_k)$ that consists of y_h (h=1, … , μ), where $d(y_h, p_b) < d(x_k, p_b)$.
7 : Set $y^* \leftarrow \mathrm{argmin}_{1 \leq h \leq \mu} f(y_h)$.
8 : If $y^* \neq p_b$, set $\mathscr{C}(p_a, p_b) \leftarrow \mathscr{C}(p_a, p_b) \cup \{y^*\}$.
9 : Set $k \leftarrow k + 1$.
10: Set $x_k \leftarrow y^*$.
11: Return to Step 5.

dMSXF necessarily moves its transition toward p_b even if all candidates in $\mathscr{N}_{\rightarrow p_b}(x_k)$ are inferior to the current solution x_k. dMSXF requires two parameters, k_{max} and μ, and $C(p_1, p_2)$ is comprised of $\{x_1, x_2, \cdots, x_{k_{max}}\}$.

Let $\mathscr{P}$ and $\mathscr{P}'$ denote the population of N_{pop} individuals and the set of the remaining individuals after sampling without replacement for the crossover, respectively. To apply dMSXF to GP, we use the following generation alternation model named CCM-Relay [9] that is familiar with local search strategies.

Procedure CCM-Relay

1 : Set $i \leftarrow 1$.
2 : Set $\mathscr{P}' \leftarrow \mathscr{P}$.
3 : Select $p_i \in \mathscr{P}'$ at random, and set $p_0 \leftarrow p_i$.
4 : Set $\mathscr{P}' \leftarrow \mathscr{P}' \setminus p_i$.
5 : If $i \leq N_{pop}$, then perform Steps 6–14, otherwise stop.
6 : If $i < N_{pop}$, then go to Step 7, otherwise set $p_j \leftarrow p_0$ and go to Step 9.
7 : Select $p_j \in \mathscr{P}'$ at random.
8 : Set $\mathscr{P}' \leftarrow \mathscr{P}' \setminus p_j$.
9 : Apply crossover to p_i and p_j.
10: Set $p^* \leftarrow \mathrm{argmin}_{1 \leq k \leq |\mathscr{C}(p_i, p_j)|} f(x_k)$, where $x_k \in \mathscr{C}(p_i, p_j)$.
11: Set $p_i \in \mathscr{P} \leftarrow p^*$.
12: Set $p_i \leftarrow p_j$.
13: Set $i \leftarrow i + 1$.
14: Return to Step 4.

3 Distance and Neighborhood Structure for Tree Structure

3.1 Definition of Locus in Tree Structure

In most discrete structures, a distance is defined based on a difference of gene at the same locus. In GPs, not only such a structural similarity but also a semantic similarity have been discussed to define the particular distance [14]. The latter one depends on problem characteristics such as a behavior of each node. Here, we consider a structural similarity to define a problem-independent distance. Since nodes are the atomic elements in a tree structure, we define the distance based on the difference of node symbols. In a tree structure, unlike in the case of most GA, the solution size is not fixed, and corresponding loci are not stable between solutions. Therefore, a kind of position adjustment between trees is required in order to define the locus, the position of node in the tree. We consider the most matching pattern of trees as a trunk to define corresponding nodes in two trees. The most matching pattern is called the largest common subgraph (subtree). Given a pair of trees, the largest common subtree (LCST, for short) is defined as the isomorphic subtree that has the maximum number of vertices between the two trees. The values stored in the nodes, i.e., symbols, do not affect whether two subtrees are isomorphic. Extracted LCST depends on the type of tree structure. Here, we treat an unordered tree that is a kind of a directed acyclic connected graph with a fixed root node and does not consider the appearance position among child node. Figure 1 shows an example for LCST.

In this figure, u_* and v_* are nodes in trees T_a and T_b. LCST is described by black solid lines. In LCST, the suffix of each node means corresponding loci between the trees and is described by $(\cdot)$. Here, the number of locus is assigned in preorder manner. The edges and nodes described by gray solid lines are parts that are exclusive to one tree, and the corresponding loci are not defined.

This definition of the locus is also introduced to the reference procedure of a probabilistic table in the probabilistic tree described in the next session.

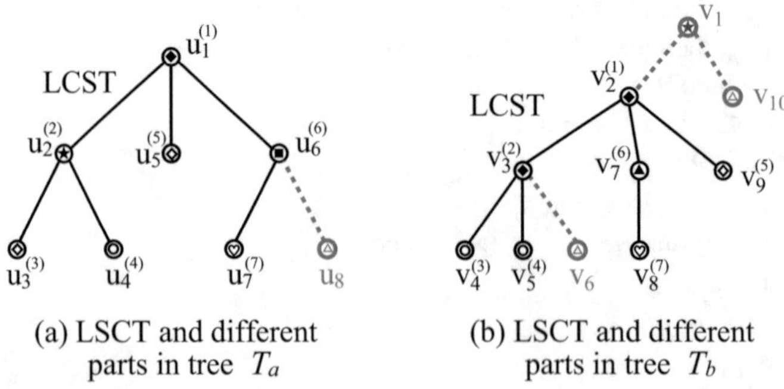

Fig. 1 LCST between trees T_a and T_b: u_1-v_2, u_2-v_3, u_3-v_4, u_4-v_5, u_5-v_9, u_6-v_7 and u_7-v_8 are corresponding loci, respectively

3.2 Definition of Distance

Let $\mathscr{U}_a$ and $\mathscr{U}_b$ denote the sets of nodes whose symbol is different from that of the corresponding node, in respective trees. In the case of Fig. 1, $\mathscr{U}_a$ and $\mathscr{U}_b$ are $\{u_2, u_3, u_6\}$ and $\{v_3, v_4, v_7\}$, respectively. The pairs of corresponding nodes u_1-v_2, u_4-v_5, u_5-v_9 and u_7-v_8 are not included in $\mathscr{U}_a$ or $\mathscr{U}_b$ because they have the same symbol each other.

Let $\mathscr{D}_a$ and $\mathscr{D}_b$ denote the sets of nodes that are not included in LCSTs, in respective trees. $\mathscr{D}_a$ and $\mathscr{D}_b$ are $\{u_8\}$ and $\{v_1, v_6, v_{10}\}$, respectively. Under the definitions, the distance between trees T_a and T_b, $d(T_a, T_b)$, is defined as

$$d(T_a, T_b) = |\mathscr{U}_a| + |\mathscr{D}_a| + |\mathscr{D}_b|, \quad (1)$$

where $|\cdot|$ means the number of components included in the set. In the case of the trees shown in Fig. 1, the distance is calculated as $3 + 1 + 3 = 7$.

3.3 Neighborhood Generation Method

In the previous work, we proposed a generation method of neighborhood candidates based on LCSTs. Our method consistently generate biased neighborhoods toward the target parent without destroying the structure of the LCST between parents. Several LCSTs exist in a tree. We extract the LCST that includes the root node tree in order to avoid missing the traits locating in the shallow place of a tree.

First, we introduce the following six functions and three fundamental operations to explain the neighborhood generation method.

3.3.1 Functions and Operations

Most problems to be solved by GP have a constraint in terms of the number of child nodes that nonterminal nodes should keep. For example, the number of operands that each arithmetic operator has is defined in the case of symbolic regression problems.

Functions are defined as follows, for describing the operators that handle tree structures.

- *parent(n)* returns the parent node of the node *n*.
- *child(n)* returns the set of child nodes of the node *n*.
- *st(n)* returns the subtree whose root node is the node *n*.
- *arg(n)* returns the number of child nodes that the node *n* should keep. It returns 0 if *n* is a terminal node.
- *symbl(n)* returns the symbol of the node *n*.
- *loc(n)* returns the locus of the node *n* if it is defined, otherwise returns *null*.

Neighborhood solutions are generated by three types of operation; *Replace*, *Delete*, *Insert*. All operations work on nodes included in the LCST.

Given two trees T_a and T_b, *Replace* is the operator that replaces the symbol of a node u of T_a with that of a node v of T_b. The operation of *Replace* is as follows.

- **Replace** (u, v)

Step 1 Select one pair of nodes u and v that satisfy $arg(u) = arg(v)$ and are located in the same gene locus, respectively from $\mathscr{U}_a$ and $\mathscr{U}_b$.
Step 2 Substitute $symbl(v)$ for $symbl(u)$ in T_a.

Delete is applied when $arg(u) > arg(v)$. This operator deletes a descendent subtree of depth 1 of the node u in T_a as follows.

- **Delete** (u, v)

Step 1 Select one pair of nodes u and v that satisfy $arg(u) > arg(v)$ and are located in the same gene locus, respectively from $\mathscr{U}_a$ and $\mathscr{U}_b$.
Step 2 Select one node included in $\mathscr{D}_a$ randomly from $child(u)$, and let the selected node denote u^*.
Step 3 Select one terminal node included in $st(u^*)$ and located in the deepest place, at random. Let the selected node denote u^{**}.
Step 4 Substitute u^{**} for $parent(u^{**})$.

Insert is the operator applied when $arg(u) < arg(v)$, which inserts nodes to the node u as descendants. The operation of *Insert* is as follows.

- **Insert** (u, v)

Step 1 Select one pair of nodes u and v that satisfy $arg(u) < arg(v)$ and are located in the same gene locus, respectively from $\mathscr{U}_a$ and $\mathscr{U}_b$.
Step 2 Generate $arg(v) - arg(u)$ terminal nodes randomly, and append to u as its child nodes.
Step 3 Substitute $symbl(v)$ for $symbl(u)$.

These operations are selected according to the number of operands of the node, exclusively. After applying the operations, the constraint in terms of the number of operands is kept.

3.3.2 Generation Method of Neighborhood Solutions

At the step k in the local search from the parent p_a to the parent p_b, neighborhood candidates in $\mathscr{N}_{\rightarrow p_b}(x_k)$ are generated by LCST of x_k and the target p_b, as below.

Procedure Neighborhood generation (at step k)

1 : Set $\mathscr{N}_{\rightarrow p_b}(x_k) \leftarrow \emptyset$.
2 : Set $h \leftarrow 1$.

3 : Calculate the node sets $\mathscr{U}_{x_k}$ and $\mathscr{U}_{p_b}$ based on LCST between x_k and p_b.
4 : If $h \leq \mu$, then perform Step 5–13, otherwise stop.
5 : Set $y_h \leftarrow x_k$.
6 : If $d(y_h, x_k) < d_{max}$, then perform Steps 7–10, otherwise perform 11.
7 : Select a node $u \in \mathscr{U}_{x_k}$ randomly and find the node $v \in \mathscr{U}_{p_b}$, where $loc(u) = loc(v)$.
8 : Apply $Replace(u, v)$, $Delete(u, v)$ or $Insert(u, v)$ to y_h according to $arg(u)$ and $arg(v)$.
9 : Recalculate $\mathscr{U}_{x_k}$ and $\mathscr{U}_{p_b}$.
10: Return to Step 6.
11: Set $\mathscr{N}_{\rightarrow p_b}(x_k) \leftarrow \mathscr{N}_{\rightarrow p_b}(x_k) \cup \{y_h\}$.
12: Set $h \leftarrow h + 1$.
13: Return to Step 4.

At step 6, d_{max} is selected randomly with the range of $[1, 2 \times d(p_a, p_b)/k_{max} - 1]$ at each step of the local search. *Replace*, *Delete* and *Insert* are applied to y_h until y_h satisfies $d(y_h, x_k) \geq d_{max}$.

4 A New Probabilistic Model-Based Neighborhood Generation

In the operation of *Delete* of the generation method of neighborhood solutions described in the previous section, the node is selected for the deletion randomly from the candidate nodes. *Insert* also appends random terminal nodes to the transitional solution in the local search to generate neighborhood solutions. Such an equal-probability random selection might sometimes cause undesirable traits that make the solution be worse.

In our study, probabilistic information obtained from previous searches of GP is used to improve the local search performance of dMSXF. To keep a appearance frequency of symbols on a node, we adopt the probabilistic prototype tree (PPT) that is one of elements used in estimation distribution algorithms for GP [4, 6]. PPT is expressed as a complete tree, and each node in PPT keeps a probability table that is relevant to all defined symbols. The symbol of a generated node is determined based on the probability table.

4.1 Previous Method

In our previous work, we applied the independent model used in Probabilistic Incremental Program Evolution (PIPE) that treats a node individually to the neighbor-

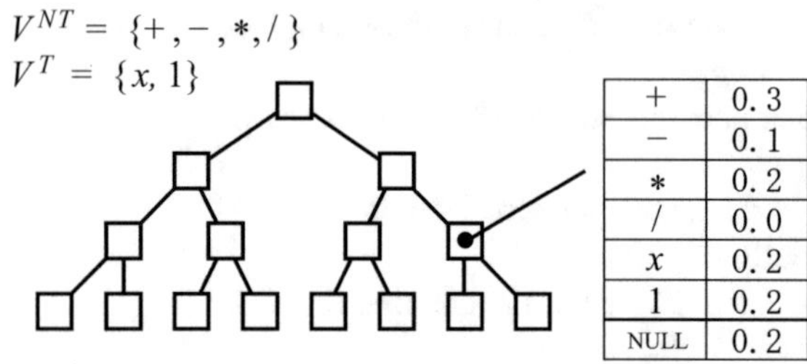

$V^{NT} = \{+, -, *, /\}$
$V^T = \{x, 1\}$
+ 0. 3
− 0. 1
∗ 0. 2
/ 0. 0
x 0. 2
1 0. 2
NULL 0. 2

Fig. 2 An example of PPT of depth 4 with independent model in the symbolic regression problem with the nonterminal set $V^{NT} = \{+, -, *, /\}$ and the terminal set $V^T = \{x, 1\}$

hood generation of dMSXF [12]. Each node has a probability table relating to ease of occurrence of symbols as shown in Fig. 2. The probability table is calculated from a frequency distribution obtained by solutions observed in the previous searches. In Fig. 2, NULL means the frequency where node has not existed at this place.

In the operation of *Insert*, a symbol with higher probability in the table is assigned frequently to a generated node. The node to be deleted is then selected in proportion to NULL in the operation of *Delete*.

4.2 Proposed Method

In many cases of tree structured problem, the dependence between nodes, especially a parent and a child, is high, which has an intensified impact on the fitness value of a solution. Here, we propose a new neighborhood generation that consider a dependency relationship between the parent node and the child node. The probability table used in Estimation of Distribution Programming (EDP) [13] is applied. Each node keeps a probability table that is relevant to symbols of parent and child as shown in Fig. 3. A symbol of node newly generated by the neighborhood generation proce-

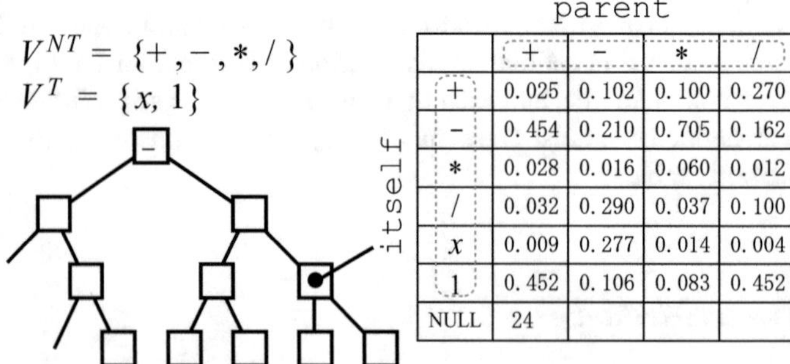

$V^{NT} = \{+, -, *, /\}$
$V^T = \{x, 1\}$
−
parent
itself
+ − ∗ /
+ 0. 025 0. 102 0. 100 0. 270
− 0. 454 0. 210 0. 705 0. 162
∗ 0. 028 0. 016 0. 060 0. 012
/ 0. 032 0. 290 0. 037 0. 100
x 0. 009 0. 277 0. 014 0. 004
1 0. 452 0. 106 0. 083 0. 452
NULL 24

Fig. 3 An example of PPT of depth 4 with parent-child model in the symbolic regression problem with the nonterminal set $V^{NT} = \{+, -, *, /\}$ and the terminal set $V^T = \{x, 1\}$

dure is determined as follows: First, find the column from the symbol of its parental node in the table. Then the symbol of the node is determined in accordance with the probabilistic distribution shown in thel column. The root node has no parents, so it does not have the probability table.

4.3 Update and Reference of PPT

For both the independent model and the parent-child model, the initial probability table of each node in PPT is set to the uniform distribution where every symbols appear with equal probability, except for NULL. When we add the information of a new tree T, i.e., a new solution, to PPT, first, the LCST of T and PPT is calculated. In accordance with LCST, the probability table of each corresponding node in PPT is then updated with the symbol (symbols) of the node of T.

In addition to updating of the probability table, the appearance frequency at each node of PPT is counted. Initial value of the appearance frequency at each node of PPT is 0. Since a solution is not necessarily the complete tree while the PPT is, the nodes that are not included in LCST exist in PPT. Therefore, the appearance frequency of only each node in the LCST increases. NULL is calculated by subtraction of the appearance frequency of the node at this place from the total number of updating.

When *Insert* is applied to a tree T, LCST of T and PPT is calculated so that the locus of a new terminal node to be inserted to T is defined on PPT. After identifying the locus of the new node, its symbol is stochastically assigned according to the probability table of the corresponding node in PPT.

In the operation of *Delete*, first, LCST of T and PPT is calculated in order to define the loci. The node of T to be deleted is then selected in proportion to the number of NULL.

4.4 Probabilistic Model-Based Neighborhood Generation

In the local search of dMSXF with the independent model or the parent-child model, the operations *Delete* and *Insert* are redefined as follows.

- **Probabilistic Model-Based Delete** (u, v)

Step 1 Select one pair of nodes u and v that satisfy $arg(u) > arg(v)$ and are located in the same gene locus, respectively from $\mathcal{U}_a$ and $\mathcal{U}_b$.
Step 2 Select one node included in $\mathcal{D}_a$ randomly from $child(u)$, and let the selected node denote u^*.
Step 3 Select one terminal node included in $st(u^*)$ and located in the deepest place, by a roulette selection relevant to the number of NULL. Let the selected node denote u^{**}.
Step 4 Substitute u^{**} for $parent(u^{**})$.

- **Probabilistic Model-Based Insert** (u, v)

Step 1 Select one pair of nodes u and v that satisfy $arg(u) < arg(v)$ and are located in the same gene locus, respectively from $\mathcal{U}_a$ and $\mathcal{U}_b$.
Step 2 Generate $arg(v) - arg(u)$ terminal nodes without assigning symbols, and append to u as its child nodes. Let the set of these new nodes be N.
Step 3 Substitute $symbl(v)$ for $symbl(u)$.
Step 4 Assign a symbol to the node included in N, in accordance with its corresponding probability table.

In the parent-child model, the frequency distribution at column of $symbl(v)$ is referred at Step 4 in *Insert*.

5 Numerical Experiments

We call dMSXF with the conventional independent model and the parent-child model are dMSXF+PIPE and dMSXF+EDP, respectively. To show the effectiveness of incorporation of the parent-child model to dMSXF, we compare dMSXF+EDP to dMSXF and dMSXF+PIPE, on the symbolic regression problem.

5.1 *Problem Domain and Instance*

Let V^{NT} and V^{T} denote the set of nonterminal nodes and the set of terminal nodes, respectively. We employed two functions as below to evaluate the search performance.

Function (I): The function

$$f = x \times \sin(\cos x - 0.5)$$

is estimated from the training set consisting of 21 data points that are placed at equal interval in the domain [0, 10]. $V^{NT} = \{+, -, \times, /, \sin, \cos\}$ and $V^{T} = \{x, 0.0, 0.1, 0.2, 0.3, 0.4, 0.5, 0.6, 0.7, 0.8, 0.9\}$ are used to express the function, where x is a variable.

Function (II): The function

$$f = x^6 + x^5 + x^4 + x^3 + x^2 + x$$

is estimated from the training set consisting of 21 data points that are placed at equal interval in the domain [−1, 1]. $V^{NT} = \{+, -, \times, /, \sin, \cos\}$ and $V^{T} = \{x, 0, 1\}$ are used to express the function, where x is a variable.

In these instances, sin and cos are unary functions that have one child nodes, and other arithmetic operators are binary functions that have two child nodes. To solve them, a binary complete tree is required in order to express the PPT.

The objective function is the sum of error at each sample point, and GP minimizes the objective function. We assume that the population reaches the optimal solution when the estimate function becomes equivalent to the defined function. In these instances, the operation that strongly degrades the objective function value of a solution is division by zero. We call such a operation the lethal trait. If a solution includes the lethal trait, its fitness value is set to ∞.

5.2 Experimental Setting

In the experiments, the population size was set to 50 and each run was terminated after 200 generations. The generation alternation model described in Sect. 2 was used. In three methods, k_{max} was set to 3 and μ was 4 for both instances. These settings were derived from preliminary experiments. The objective of the experiments is to examine the effectiveness of crossovers, therefore neither mutations nor bloat controlling strategies were applied. Initial solutions of GPs were generated randomly, where every symbols to be assigned to each node were selected with equal probability.

An initial tree, a solution, was generated with a size in the range of [11, 25] or [25, 35] nodes. The depth of PPT was set to 8. PPT was updated with upper half of the neighborhood solutions generated in every local searches in terms of the objective function value.

We conducted 50 runs, and PPT was initialized in conjunction with the initialization of the population in the first run. After the second run, updated PPT of the previous run were carried over to the next.

5.3 Comparison Results

Table 1 shows the number of trials that reach the optimum (#opt), and the average of the number of the evaluations to find the optimum (#eval). In Table 2, the average of the number of the nodes constituting the obtained optimal trees, and the average depth of the trees are shown.

Figure 4 illustrates the transition of the average of objective function value of the best solutions out of 50 runs. This result was obtained by dMSXFs whose initial trees were generated with a size in the range of 25–35, on Function (I).

As shown in Table 1, regardless of the setting in the initial tree size, dMSXF+EDP and dMSXF+PIPE can reach the optimal solution with high probability in the two instances. From Table 2, no trend in both the size and the depth of the trees obtained by dMSXFs is found. In most case, dMSXF+EDP is found

Table 1 Comparison of the search performance

Function	Inital size	dMSXF		dMSXF+PIPE		dMSXF+EDP	
		%opt	#eval	%opt	#eval	%opt	#eval
I	11–21	0.94	2.5×10^4	**1.00**	$\mathbf{1.6 \times 10^4}$	0.98	1.7×10^4
	25–35	0.90	3.7×10^4	0.96	2.4×10^4	**1.00**	$\mathbf{2.0 \times 10^4}$
II	11–21	0.02	6.1×10^4	0.16	2.4×10^4	**0.26**	$\mathbf{2.4 \times 10^4}$
	25–35	0.46	3.5×10^4	0.68	1.7×10^4	**0.80**	$\mathbf{1.5 \times 10^4}$

Table 2 Comparison of the tree size

Function	Inital size	dMSXF		dMSXF+PIPE		dMSXF+EDP	
		#size	#depth	#size	#depth	#size	#depth
I	11–21	13.0	4.5	12.6	4.5	**11.5**	**4.3**
	25–35	21.2	6.1	**16.0**	**5.3**	17.0	5.4
II	11–21	**14.9**	**4.9**	15.3	5.1	15.6	5.1
	25–35	20.6	5.9	**18.4**	**5.8**	20.0	6.1

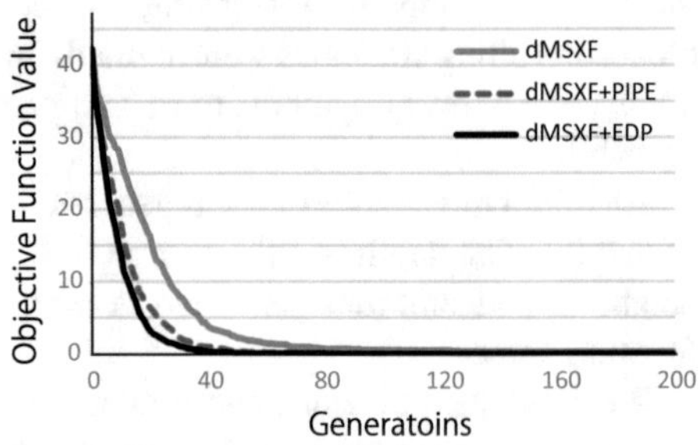

Fig. 4 Transition of objective function value of dMSXFs with the large initial tree, on Function (I)

to be superior to dMSXF and dMSXF+PIPE. Furthermore, we can see that EDP makes the search more effectively since dMSXF+EDP can find the optimum with low evaluation cost. Its faster convergence can be also observed in the transition of the objective function value of the best fitness, as shown in Fig. 4.

Figure 5 shows the average of lethal traits included in one solution. This result was obtained by dMSXFs whose initial trees were generated with a size in the range of 25–35. The average was calculated from all evaluated solution in all runs of dMSXFs.

From Fig. 5, we can see that probabilistic models succeed in suppressing the generation of the lethal trait even though they do not explicitly prohibit them. Remarkable difference between dMSXF+PIPE and dMSXF+EDP can be found in Function (II).

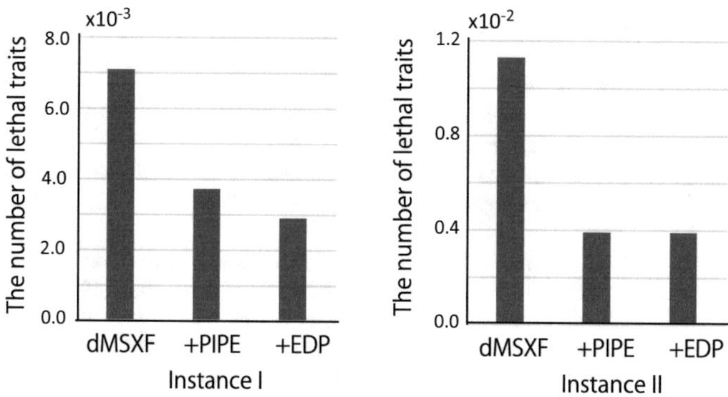

Fig. 5 Average of the lethal trait (division by 0) included in one solution

6 Conclusion

In this paper, we proposed a new neighborhood generation that considers a parent-child dependency relationship for dMSXF, in order to improve efficiency of both heredity and composition of preferable traits. Through numerical experiments with a symbolic regression problem, it was shown that our new method finds optimal solutions with high probability compared to the conventional probabilistic dMSXF. The convergence speed was improved by the proposed method. In addition, it was shown that the generation of lethal traits is suppressed by using probabilistic information found in preferable solutions. Further improvement is expected by applying more sophisticated approaches such as Bayesian network to treat the complex dependency relation between nodes. In the processes of both update and reference of the probabilistic prototype tree, LCST is calculated, which calls more calculation time as compared to the original dMSXF. That problem in runtime can be resolved by a parallelization technique. These tasks are left as a future goal.

Acknowledgements This work was supported by JSPS KAKENHI Grant Number 26330290.

References

1. Langdon, W.B.: Size fair and homologous tree crossovers for tree genetic programming. Genet. Program. Evolv. Mach. **1**(112), 95–119 (2000)
2. Keijzer, M., Ryan, C., O'Neill, M., Cattolico, M., Babovic, V.: Ripple crossover in genetic programming. In: Proceedings of the 4th European Conference on Genetic Programming (EuroGP). Lecture Notes in Computer Science, vol. 2038, pp. 74–86 (2001)
3. Semenkin, E., Semenkina, M.: Self-configuring genetic programming algorithm with modified uniform crossover. In: IEEE World Congress on Computational Intelligence (WCCI) 2012, pp. 1–6 (2012)
4. Sałustowicz, R.P., Schmidhuber, J.: Probabilistic incremental program evolution. Evolut. Comput. **5**(2), 123–141 (1997)

5. Majeed, H., Ryan, C.: A less destructive, context-aware crossover operator for GP. In: Proceedings of the 9th European Conference on Genetic Programming (EuroGP). Lecture Notes in Computer Science, vol. 3905, pp. 36–48 (2006)
6. Hasegawa, Y., Iba, H.: A Bayesian network approach to program generation. In: Proceedings of IEEE Transactions Evolutionary Computation **12**(6), 750–764 (2008)
7. Beadle, L., Johnson, C.G.: Semantically driven crossover in genetic programming. In: Proceedings of IEEE World Congress on Computational Intelligence (WCCI) 2008, pp. 111–116 (2008)
8. Yamada, T., Nakano, R.: Scheduling by genetic local search with multi-step crossover. In: Proceedings of Parallel Problem Solving from Nature IV, pp. 960–969 (1996)
9. Ikeda, K., Kobayashi, S.: Deterministic multi-step crossover fusion: a handy crossover for GAs. In: Proceedings of Parallel Problem Solving from Nature VII, pp. 162–171 (2002)
10. Hanada, Y., Hiroyasu, T., Miki, M.: Genetic multi-step search in interpolation and extrapolation domain. In: Proceedings of Genetic and Evolutionary Computation Conference 2007, pp. 1242–1249 (2007)
11. Hanada, Y., Hosokawa, N., Ono, K., Muneyasu, M.: Effectiveness of multi-step crossover fusions in genetic programming. In: Proceedings of IEEE Congress on Evolutionary Computation, pp. 1743–1750 (2012)
12. Matsumura, K., Hanada, Y., Ono, K.: Probabilistic model-based multistep crossover for genetic programming. In: Proceedings of 2016 Joint 8th International Conference on Soft Computing and Intelligent Systems 2016, pp. 154–159 (2016)
13. Yanai, K., Iba, H.: Estimation of distribution programming based on Bayesian network. In: Proceedings of the Congress on Evolutionary Computation 2003, pp. 1618–1625 (2003)
14. Riolo, R., Worzel, B., Kotanchek, M.: Genetic programming theory and practice XII, Springer International Publishing (2015)

Digital Watermark Design for Two-Dimensional Codes Displayed on Smart Phone Screen Using Multi-objective Optimization and Optical Simulation

Shingo Takeshita, Takeru Maehara and Satoshi Ono

Abstract A two-dimensional (2D) code has been used for authentication in train and airplane boarding passes, e-commerce etc. However, inappropriate use of copied 2D code is apprehended because 2D code can be easily duplicated. Therefore, this study proposes a method for designing a digital watermark that detects replication of 2D code displayed on a smartphone screen. To achieve this, the proposed method designs an effective watermarking scheme for various smartphone models using multi-objective optimization including optical simulation. Experiments showed that the proposed method can design a watermarking scheme with the same performance as the previous work whereas it does not require using actual smartphone devices during long-time optimization.

1 Introduction

Barcodes are used for object recognition and identification in various areas, such as production, logistics, and commerce. Two-dimensional barcodes (2D codes) are currently used as a shorthand method (an "analogue shortcut") to input, for example, a URL, an e-mail address, a phone number, and so on. In recent years, 2D codes have also been used for authenticating airplane boarding passes and online payments. In particular, 2D codes displayed on smart phone screens have become increasingly common as a form of paperless verification. However, the 2D codes can easily be replicated by malicious replication or fabrication. These problems will become more serious with the popularization of mobiles and online payment services using 2D codes.

S. Takeshita (✉) · T. Maehara · S. Ono
Department of Information Science and Biomedical Engineering, Graduate School of Science and Engineering, Kagoshima University, Kagoshima 890-0065, Japan
e-mail: sc113035@ibe.kagoshima-u.ac.jp

S. Ono
e-mail: ono@ibe.kagoshima-u.ac.jp

R. Lee (ed.), *Software Engineering, Artificial Intelligence, Networking and Parallel/Distributed Computing*, Studies in Computational Intelligence 721,
DOI 10.1007/978-3-319-62048-0_14

Fig. 1 Malicious replication using smart phones

Recently, digital watermarking has been used widely to protect copyrights and detect image modification and attacks. Digital watermarking technology can roughly be classified as robust, fragile or semi-fragile ones. Semi-fragile watermark has the characteristics that it is destroyed when subjected to certain processing. Compared to robust and fragile watermark, little attention has been paid to semi-fragile watermark [1, 2].

Ono et al. proposed a semi-fragile watermarking scheme for color 2D barcodes that detects malicious replication [3–6]. As shown in Fig. 1, the duplication targeted in the research is to capture a 2D code displayed on a smartphone with another smartphone's camera. Such easy rephotographing allows replication of 2D codes. They proposed a method for detecting duplicated 2D codes by embedding semi-fragile watermark into the code. The watermark can be extracted from the regular 2D code (original), whereas it is not extracted from the duplicated code since it is destroyed by rephotographing (duplication). Therefore, identification of the original and duplication becomes possible by presence/absence of the watermark. In this method, an automatic design method of semi-fragile digital watermark for 2D code is realized by using evolutionary multi-objective optimization. However, in order to evaluate solution candidates, actual smartphones must be used throughout the long hours of optimization. The previous work requires totally $2n^2$ devices where n models are used.

Therefore, this study attempts to introduce an optical simulation to Ono's optimization-based watermark design method [5, 7] for evaluating solution candidates. In the proposed simulation, the optical transfer functions (OTFs) of a smartphone screen and a camera, interference pattern between them, and noise are observed before the optimization. Then, photographing and duplication (rephotographing) of watermarked 2D codes are performed virtually by the simulation, and the watermark is extracted from the estimated images when evaluating solution candidates. As a result, the actual smartphone devices are necessary only when constructing the simulator beforehand, and the devices become unnecessary during the optimization. Also, regardless of the total number of smartphone models used in the optimization, the number of required devices is limited only two for each model.

Experiments showed that the proposed method can design a semi-fragile watermark with performance equivalent to that of the previous work using actual devices for solution evaluation.

2 Related Work

Watermark design by optimization has been extensively studied. Vahedi et al. proposed a method for embedding digital watermarks to color images using Discrete Wavelet Transform (DWT) [8, 9]. Mingzhi et al. proposed an optimization method of digital watermark embedding strength for the frequency bands obtained by DWT and discrete cosine transform using genetic algorithm [10].

Research for optical simulations has also been widely conducted to simulate photographing by a digital camera. Farrell et al. proposed a simulator for design and performance evaluation of digital cameras [11]. This method models light sources of a scene to be photographed, photons passing through lenses, sensor response to colors, etc. They developed a software library Image System Evaluation Toolbox (ISET) using their simulator. Chen et al. built a more rigorous simulator by extending the ISET simulator [12] so that an influence caused by a crosstalk between light rays incident on a sensor is estimated.

3 Proposed Method

3.1 Key Ideas

In this study, we propose a digital watermark design method based on the following idea.

Formulation as a Multi-objective Optimization Problem:

In the proposed method, an Objective function represents semi-fragileness of the designed watermarking scheme. Because the semi-fragileness may depend on display types such as liquid crystals (LCD) and organic Electro Luminescence (EL), the semi-fragileness is calculated for each smartphone model combinations, i.e., one for displaying a watermarked 2D code and another for rephotographing it. In addition, as with the earlier study [5, 7], in order to design a digital watermark which can be commonly versatile for various smartphones, the proposed method designs a watermarking scheme using multi-objective optimization rather than aggregating the objective functions. This is because, generally, a linear weighted sum of multiple objective functions in problems whose objective functions have strong trade-off or are ill-balanced, it might prioritize one of the functions, i.e., watermarking schemes inappropriately prioritizing one smartphone model might be designed. On the other hand, using the worst value of the functions as a fitness to aggregate the functions into single one might prevent individuals improved in only one objective from being selected to the next generation and contributing the optimization [13].

The multi-objective optimization makes it possible not only to design the semi-fragile watermarking scheme that is commonly useful for various smartphone models but also to obtain knowledge such as effective watermark components for different smartphone models.

An Optical Simulation for Solution Candidate Evaluation:

In the case that replication between different smartphone models is considered as well as that between the same model when evaluating solutions (candidate watermarking schemes), $2n$ actual devices for each model are necessary in the previous method [5, 7]. Particularly to EL screen, a residual image occurs during long-run optimization.

Therefore, the proposed method adopts an optical simulation to evaluate the semi-fragileness of the candidate watermarking scheme, where photographing and replicating (rephotorgraphing) watermarked 2D code images are performed approximately. Since semi-fragile watermarks are mainly embedded in high-frequency components of a cover image, the proposed simulator places importance on simulation of phenomena affecting high-frequency components.

3.2 *Formulation*

3.2.1 Design Variable

The proposed method simultaneously selects frequency bands used for watermark embedding and determines its embedding strength within an optimization. The frequency subband selection and strength adjustment are performed for each cover image region; as shown in the Fig. 2, the cover image is divided into three regions of a dark module region D, a light module region B, and an edge region E based on luminance values. Also, in the proposed method, two levels of DWT are applied to the image to be embedded, as shown in Fig. 3, obtaining seven subbands {HH1, HL1, LH1, HH2, HL2, LH2, LL2}. Totally, the number of combinations of subband and image region is 21, which corresponds to the number of design variables in the optimization.

Design variables corresponding to the image region r and the frequency band b are denoted by $v_{r,b}$, and real numbers from 0 to 1 are taken as its value. The value simultaneously represents information for subband selection and embedding strength adjustment, i.e., the watermark is embedded to region r in subband b with the strength $L_{r,b}$:

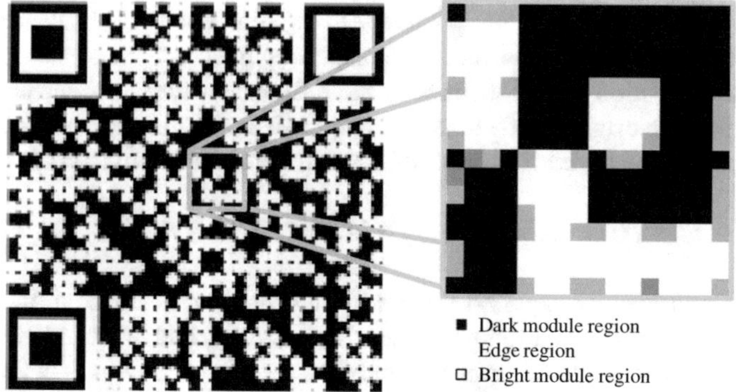

Fig. 2 Three image regions in cover 2D code in which different watermarking schemes are used

Fig. 3 Subbands obtained by two-level DWT

$$L_{r,b} = 2 \times \left(v_{r,b} - 0.5\right) \times L_{max}, \tag{1}$$

where L_{max} represents the maximum strength. In other words, the watermark is embedded only when the value of $v_{r,b}$ exceeds the threshold 0.5, which was determined so that 50% of the subbands would be used when generating initial solutions.

3.2.2 Objective Function

Objective functions in the proposed method are calculated from semi-fragileness of a candidate watermarking scheme, which should be maximized. Since the proposed method performs multi-objective optimization rather than using single objective function, each objective function represents semi-fragileness for each device model combination, i.e., a set of smartphone models for representing watermarked 2D code and for capturing it. For example, when designing a watermark commonly useful for two models M_1 and M_2, the number of combinations is four: (M_1, M_1),

(M_2, M_2), (M_1, M_2), and (M_2, M_1), where (M_i, M_j) represents that M_i is used for displaying watermarked 2D code and M_j for capturing it.

Objective function $f_k(i)$ is defined by the difference between watermark extraction rates from an original code image and replicated one. The watermark extraction rate is expressed by bit correct ratio (BCR), and the objective function is defined as follows.

$$f(i) = BCR(W, W^{vld}) - BCR(W, W^{rpl}) - P(Y^{vld}) \tag{2}$$

where Y^{vld} and Y^{rpl} are captured images of an original watermarked 2D code and a replica displayed on smartphone screens respectively, and W^{vld} and W^{rpl} are the watermark images extracted from Y^{vld} and Y^{rpl}, respectively. BCR is defined by the following equation:

$$BCR\left(W, W'\right) = 1 - \frac{\sum_{w=1}^{w_W} \sum_{h=1}^{h_W} (W_{w,h} \oplus W'_{w,h})}{w_W h_W}, \tag{3}$$

where $W_{w,h}$, $W'_{w,h}$ are the values of the pixels located at (w, h) coordinates of the embedded watermark image W and the extracted watermark image W', $h_W \times w_W$ represents the size of the watermark image, and $\oplus$ is exclusive OR operator, respectively. We introduce the penalty function P defined by the following equation.

$$P\left(Y^{vld}\right) = ECR\left(Y^{vld}\right) \times P_{max}, \tag{4}$$

where ECR represents error correction function usage rate used when decoding the 2D codes of the cover image in the photographed image of the original. $0 \leq ECR\left(Y^{vld}\right) \leq 1$.

3.3 Process Flow

As shown in Fig. 4, the process flow of watermarking scheme design by the proposed method is basically the same as the previous work [5] except the evaluation process; the optical simulator is used to evaluate solution candidates whereas the previous work uses actual smartphone devices. Although various evolutionary multi-objective optimization algorithms can be used in the proposed method, this study uses Nondominated Sorting Genetic Algorithm II (NSGA-II) [14] because its behavior is well-known and allows us to easily analyze optimization result. Similar to general population-based meta-heuristics, NSGA-II iterates solution candidate reproduction and their evaluation.

Evaluation for solution candidates are performed as follows. To calculate Eq. (2), first, a watermark image is embedded to a cover 2D code image according to the watermarking scheme (corresponding solution candidate). Then, the watermarked 2D code image displayed on a smartphone screen is virtually captured as described

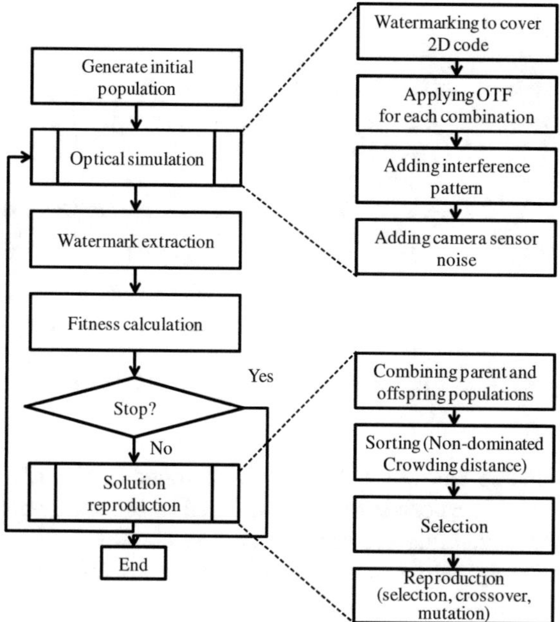

Fig. 4 Process flow of the proposed method

in Sect. 3.4, and its watermark is extracted by applying DWT and combining frequency components of the subbands involving the watermark with weights corresponding the watermarking strength levels. Next, replication of the watermarked 2D code image is virtually captured, and its watermark is extracted in the same way.

3.4 *Watermark Evaluation Simulation*

The proposed method simulates photographing and rephotographing (replication) of watermarked 2D codes rather than using actual devices as described in Sect. 3.1. In this simulation, phenomena affecting high frequency components of images must be estimated since watermarking schemes designed by the proposed method mainly embeds the watermark into the high frequency components to achieve semi-fragileness. Therefore, the proposed method estimates photographed and rephotographed images by calculating OTF of cameras and displays, interference patterns between a camera and a screen, and sensor noises of a camera, which are measured before the optimization. The overview of the proposed simulation is shown in Fig. 5 and below.

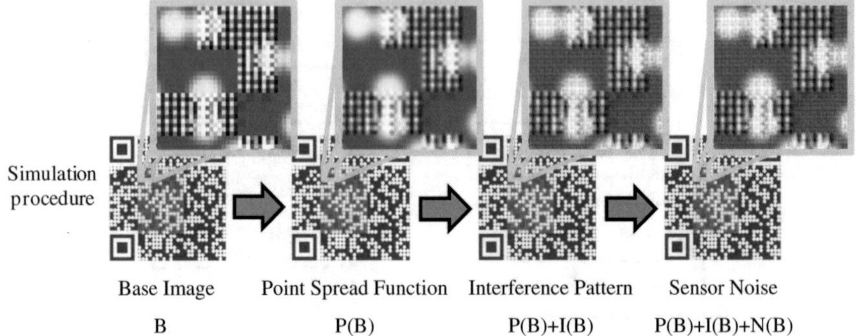

Fig. 5 Procedure of the proposed optical simulation and example images

Step 1: (Reproducing blur effect) In order to reproduce a blur of a target camera, convolution is calculated with an original image to be photographed with the point spread function of the camera as a kernel.

Step 2: (Reproducing interference pattern) Add the interference pattern obtained for each color to the target image.

Step 3: (Reproducing camera sensor noise) Add the camera sensor noise to the target image.

The above simulation is formulated as follows:

$$S(B) = P(B) + I(B) + N(B), \tag{5}$$

where B and $S(B)$ represent the original image (watermarked 2D code image) and its estimated image, and $P(\cdot)$, $I(\cdot)$, and $N(\cdot)$ correspond to the optical transfer function, the interference pattern, and the camera sensor noise, respectively. Rephotographing (replication) is also estimated in the same way except that an original watermarked 2D code image is replaced with its virtually photographed image. In the proposed simulation, a blur effect by photographing is represented by a point spread function (PSF). The PSF of a combination of a camera and a screen is obtained by the following procedure: first, a single pixel is displayed on a target smartphone display, which is regarded as a point light source, and then, a target camera captures the pixel. We do this with all the combinations of cameras and smartphone screens.

3.4.1 Interference Pattern

In general, when capturing a display with a digital camera, periodic patterns (interference patterns) occur on a captured image. This pattern needs to be estimated since it affects watermark extraction. Therefore, the proposed method acquires the interference patterns before optimization. Since it varies depending on a color as well as a pair of camera and screen, the pattern is defined for each color by capturing flat images having only one color. Therefore, the interference pattern $I(\cdot)$ to be obtained can be expressed by the following equation.

$$I(B) = \sum_{col \in B}^{N_c} I^{(col)}(B), \tag{6}$$

where B is a target image, col is one of colors comprising B, and N_c is the number of all colors comprising B, respectively.

The interference pattern acquisition procedure is shown below.

Step 1: (Photographing) A flat image colored with only one color col is photographed and an obtained image is denoted as J_{col}.
Step 2: (Averaging brightness) From the captured image J_{col} obtained in Step 1, average brightness value b_{col} is obtained for each for channel.
Step 3: (Pattern acquisition) A target interference pattern $I^{(col)}$ is obtained by subtracting b_{col} from each pixel of J_{col}.

Camera sensor noise is observed as high frequency noise, which affects the watermark. In general, sensor noise can be classified into photon electron shot noise, dark current shot noise, read noise, fixed pattern noise [15]. Since it is assumed that the above noise occurs according to the Poisson distribution or Gaussian distribution, their variance can be approximated from the captured images. Healey et al. proposed a method to obtain the noise variances by photographing several times [16], and the proposed method also estimates the variances in the same manner. The pattern $N(B)$ representing the camera sensor noise consists of the expectation value $D(a, b)$ of the luminance value at the coordinates (a, b) as follows:

$$\begin{aligned} D(a,b) &= (K(a,b)I(a,b) + N_{DC}(a,b) \\ &\quad + N_S(a,b) + N_R(a,b))A + N_Q(a,b) \end{aligned} \tag{7}$$

where $I(a, b)$, $K(a, b)$, $N_{DC}(a, b)$, $N_S(a, b)$, $N_R(a, b)$, $N_Q(a, b)$ and A are brightness of B, fixed pattern noise, dark current shot noise, photon electron shot noise, read noise, quantization noise, and gain, respectively. In the proposed method, each of the above noise components was acquired individually.

4 Evaluation

4.1 Experimental Settings

To verify the effectiveness of the proposed method, we tried to design semi-fragile watermarking schemes commonly useful for two smartphone models, which have LED and EL, respectively, i.e., the representative two display types used in smartphones on the market:

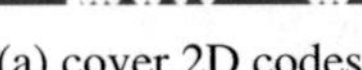

(a) cover 2D codes

(b) Watermark image

Fig. 6 The cover 2D code and watermark image used in this experiment

- D_1: SHARP AQUOS PHONE ZETA SH-06E (Display: 4.8 in. IGZO, 1,920 × 1,080 pixels, camera: 13 million pixels)
- D_2: Galaxy S6 SC-04E (Display: 5.0 in. Full HD Super AMOLED, 1,920 × 1,080 pixels, camera: 13 million pixels)

The digital camera used to extract the watermark is as follows:

- CCD camera: Pointgrey FLEA3 FL3-U3-88S2C-C, 4,092 × 2,160 pixels
- Lens: FUJIFILM DV3.4 × 3.8SA-1

In the experiment of this paper, replication using different two smartphone models, e.g., taking a 2D code displayed on the screen of D_1 with the camera of D_2, is performed as well as replication using the same model, whereas the previous work performed replication using the same model only. Therefore, the number of objective functions was four. Figure 6 shows the cover 2D code and the embedded watermark image. In order to avoid the influence of the distortion caused by the camera lens, the watermark is embedded in the center of the 2D code. The maximum embedding strength L_{max} was set to 192, and P_{max} was set to 2.

The crossover used for NSGA-II was blend crossover (BLX-α) [17], and the mutation used Uniform Mutation. α was 0.5, the mutation rate was 0.05, the population number was 1000, and the termination condition was when the number of generations reached 335.

4.2 Designed Solution

Figure 7 shows the non-dominated solutions obtained by the proposed method. In order to represent four objective functions, the X, Y, Z axis and shade of marker color are used, which corresponds to D_1–D_1, D_2–D_2, D_1–D_2 and D_2–D_1, respectively. We can see that solutions reaching the fitness of 0.25 were generated for all four objective functions. Figure 8 shows design variable values and watermarked 2D codes of the non-dominated solution that have the highest average of the four objective function values among the solutions having 0.25 or higher values of all the four functions. In this solution, watermarks are embedded in all HH1 components of the three regions, LH1 of the dark and light module region, HL1 of the dark module and edge regions,

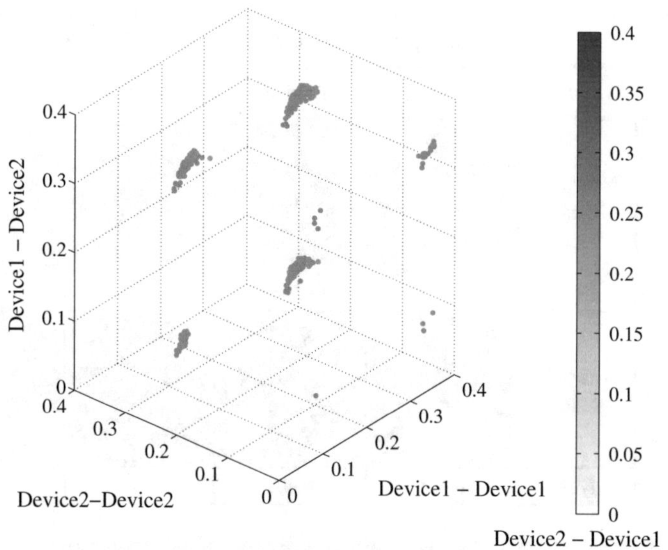

Fig. 7 Distribution of obtained non-dominated solutions in objective space

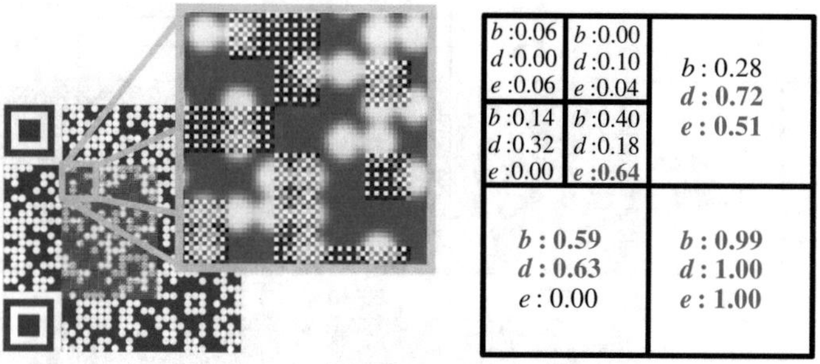

b :0.06 d :0.00 e :0.06
b :0.00 d :0.10 e :0.04
b :0.14 d :0.32 e :0.00
b :0.40 d :0.18 e :0.64
b : 0.28 d : 0.72 e : 0.51
b : 0.59 d : 0.63 e : 0.00
b : 0.99 d : 1.00 e : 1.00

Fig. 8 Semi-fragile watermark scheme designed by the optimization

and HH2 of edge regions. The difference of subband selection results between cover image regions allows realizing common effectiveness on the two smartphone models. A BCR difference of 0.25 is sufficient for 2D code authentication according to the findings obtained in the previous work [13].

The results by the proposed method and previous work that performs single objective optimization whose objective function using only D_1 were compared. Figure 9 shows the results: the extracted watermark images and BCR values from original images shown on the screens of D_1 and D_2 and replicated images by the same model. The watermark designed by the proposed method could be extracted from the original images with high BCR values whereas BCR value of the watermark designed by the single-objective optimization was lower on D_2 since D_2 was not used during optimization.

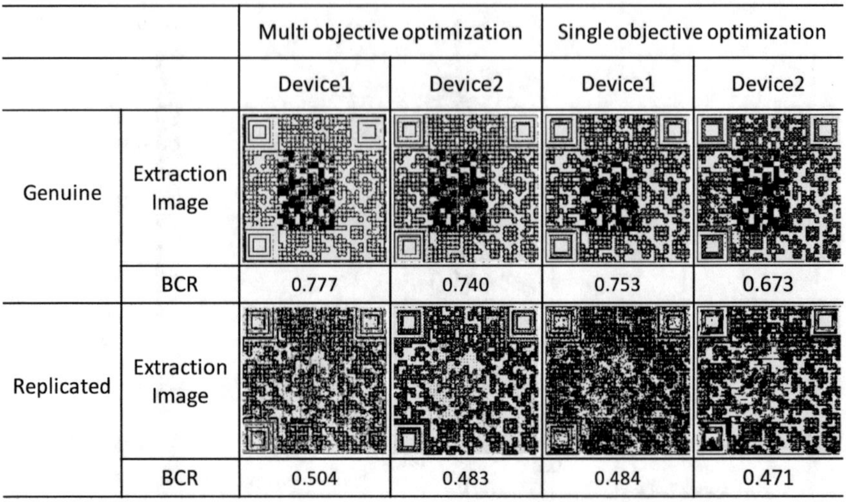

		Multi objective optimization		Single objective optimization	
		Device1	Device2	Device1	Device2
Genuine	Extraction Image				
	BCR	0.777	0.740	0.753	0.673
Replicated	Extraction Image				
	BCR	0.504	0.483	0.484	0.471

Fig. 9 Multi-objective optimization and single objective optimization comparison

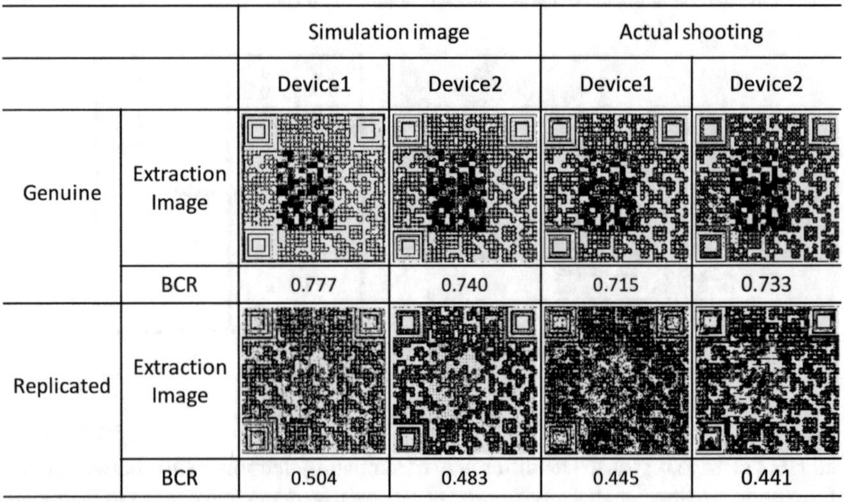

		Simulation image		Actual shooting	
		Device1	Device2	Device1	Device2
Genuine	Extraction Image				
	BCR	0.777	0.740	0.715	0.733
Replicated	Extraction Image				
	BCR	0.504	0.483	0.445	0.441

Fig. 10 Comparison of actual shooting and reproduction image

Next, to see the appropriateness of the proposed simulator, we compared the watermarks extracted from the actually captured and estimated images. Figure 10 shows the results. Similar watermark images were successfully estimated from original images though peripheral area outside the watermarked region looks different. Difference also could be seen in the result from the replicated images the estimated results did not involve almost no watermark components, whereas the watermarks were not completely destroyed but its patterns were changed in the actually captured results.

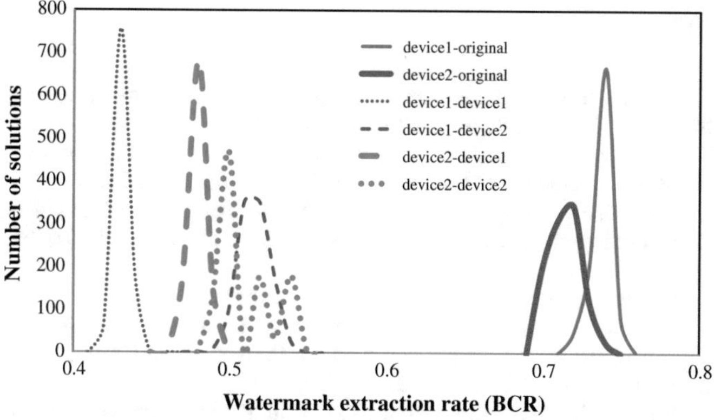

Fig. 11 Distribution of watermark extraction rate of original and reproduction

Finally, we examine how stable the authenticity can be determined with the watermark designed by the proposed method in the actual environment.

Figure 11 shows the histogram of BCR values when extracted from original and replicated images with slightly varying the extraction area in the images. From Fig. 11, the worst BCR value in the original and the best BCR value in replication were 0.69 and 0.54, respectively, resulting in the margin of 0.15. Therefore, a threshold can be easily determined for authenticity.

5 Conclusions

We propose a method for automatic design of semi-fragile watermark, aiming at detecting illegal duplication of 2D code. The proposed method simultaneously performs subband selection and strength adjustment using the optical simulation and evolutionary multi-objective optimization algorithm. Experimental results have shown that the proposed simulator appropriately estimates the watermarking and extraction result and that the proposed method designs semi-fragile watermark commonly effective to more than one smartphone models without actual smartphone devices during long hours of optimization.

On the other hand, there is room for improvement of the performance of multi-objective optimization since Pareto ranking in NSGA-II might not work properly for this problem involving four objective functions. In the future, we plan to reduce the number of objective functions; following the previous work in robust optimization with bi-objective optimization [18], it is possible to aggregate many objective functions into two functions corresponding to average and standard deviation of the function values.

References

1. Rey, C., Dugelay, J.-L.: A survey of watermarking algorithms for image authentication. EURASIP J. Adv. Signal Process. **2002**(6), 218932 (2002)
2. Song, Y.J., Liu, R.Z., Tan, T.N.: Digital watermarking for forgery detection in printed materials. In: Pacific-Rim Conference on Multimedia, pp. 403–410. Springer (2001)
3. Ono, S., Tsutsumi, M., Nakayama, S.: A copy detection method for colored two-dimensional code using digital watermarking. IEICE Trans. Inf. Syst. (Japanese ed.) **94**(12), 1971–1974 (2011)
4. Ono, S., Maehara, T., Sakaguchi, H., Taniyama, D., Ikeda, R., Nakayama, S.: Self-adaptive niching differential evolution and its application to semi-fragile watermarking for two-dimensional barcodes on mobile phone screen. In: Proceedings of Genetic and Evolutionary Computation, pp. 189–190. ACM (2013)
5. Ono, S., Maehara, T., Nakai, K., Ikeda, R., Taniguchi, K.: Semi-fragile watermark design for detecting illegal two-dimensional barcodes by evolutionary multi-objective optimization. In: Proceedings of Conference on Genetic and Evolutionary Computation, Companion, pp. 175–176. ACM (2014)
6. Ono, S., Maehara, T., Minami, K.: Coevolutionary design of a watermark embedding scheme and an extraction algorithm for detecting replicated two-dimensional barcodes. Appl. Soft. Comput. **46**, 991–1007 (2016)
7. Maehara, T., Nakai, K., Ikeda, R., Taniguchi, K., Ono, S.: Watermark design of two-dimensional barcodes on mobile phone display by evolutionary multi-objective optimization. In: Soft Computing and Intelligent Systems, pp. 149–154. IEEE (2014)
8. Vahedi, E., Zoroofi, R.A., Shiva, M.: Toward a new wavelet-based watermarking approach for color images using bio-inspired optimization principles. Digit. Signal Process. **22**(1), 153–162 (2012)
9. Kundur, D., Hatzinakos, D.: A robust digital image watermarking method using wavelet-based fusion. Proc. Intl. Conf. Image Process. **1**, 544–547 (1997)
10. Mingzhi, C., Yan, L., Yajian, Z., Min, L.: A combined DWT and DCT watermarking scheme optimized using genetic algorithm. J. Multimedia **8**(3), 299–305 (2013)
11. Farrell, J.E., Xiao, F., et al.: A simulation tool for evaluating digital camera image quality. Electron. Imaging **2004**, 124–131 (2003)
12. Chen, J., Venkataraman, K., et al.: Digital camera imaging system simulation. IEEE Trans. Electron. Devices **56**(11), 2496–2505 (2009)
13. Maehara, T., Nakai, K., Ikeda, R., Taniguchi, K., Ono, S.: Watermark design for replication detection of two-dimensional barcodes by evolutionary multiobjective optimization. IEICE Trans. Inf. Syst. (Japanese ed.) **98**(5), 835–846 (2015)
14. Deb, K., Pratap, A., Agarwal, S., Meyarivan, T.: A fast and elitist multiobjective genetic algorithm: NSGA-II. IEEE Trans. Evol. Comput. **6**(2), 182–197 (2002)
15. Nakamura, J.: Image Sensors and Signal Processing for Digital Still Cameras. CRC Press (2016)
16. Healey, G.E., Kondepudy, R.: Radiometric CCD camera calibration and noise estimation. IEEE Trans. Pattern Anal. Mach. Intell. **16**(3), 267–276 (1994)
17. Eshelman, L.J.: Chapter real-coded genetic algorithms and interval-schemata. Found. Genet. Algorithms **2**, 187–202 (1993)
18. Shimoyama, K., Oyama, A., Fujii, K.: A new efficient and useful robust optimization approach-design for multi-objective six sigma. IEEE Congr. Evol. Comput. **1**, 950–957 (2005)

Author Index

R. Lee (ed.), *Software Engineering, Artificial Intelligence, Networking and Parallel/Distributed Computing*, Studies in Computational Intelligence 721,
DOI 10.1007/978-3-319-62048-0

Printed in the United States
By Bookmasters